ELECTRICAL ENGINEERING

Theory and Examples

Fifth Edition

Published by HRC Publishers, Wayne, Pennsylvania, USA. hrcpublishers@comcast.net

ISBN 978-0-9772484-4-5

Printed in USA

ELECTRICAL ENGINEERING
Theory and Examples

Fifth Edition

K. H. Norian

Electrical and Computer Engineering Department, Lehigh University.

To my parents

Contents

Preface

The book covers laws and methods in electrical engineering for students taking the subject for the first time. It is based on foundation courses in electrical engineering that I have taught to electrical engineers, as well as to students from other engineering disciplines and the sciences, for thirty years. It is suitable as a textbook for a one semester introductory course in electrical engineering, for preparation for exams, or for self study. The text contains the circuits and the theory of the experiments that are carried out in the electrical lab courses that I have given for quite a number of years. The prerequisite for a course that uses this text is a basic knowledge of algebra and calculus.

My emphasis in writing the text has been on being clear. My aim has been to impart an intuitive understanding of the subject. The laws and methods of the first chapter set the foundation for the book and are used throughout the text. Theory is followed by solved examples drawn from past exam questions.

K. H. Norian

1.1 Introduction

The application of electrical energy in useful systems is implemented through the use of electrical circuits. Electrical circuits are made up of circuit elements connected together, with conductors, in closed loops. Conductors, resistors, inductors, capacitors, and voltage and current sources are circuit elements. A conductor is a wire in a physical circuit and is represented by a straight line in a circuit diagram. A conductor is assumed to have zero resistance to the flow of current. All points in a circuit that are connected by conductors represent a single node. The terminals of a resistor are two distinct nodes. A resistor connected between two nodes forms a branch. Connecting two points in a circuit with a conductor results in a short circuit between the two points. If two points in a circuit are not connected by a conductor, or any other circuit element, then an open circuit exists between the two points. Voltage sources and current sources cause currents to flow and voltage differences to appear in the circuit. This results in the delivery of electrical power to different parts of the circuit. The analysis of circuits to determine currents, voltages and power involves the use of methods based on electrical laws that will be applied in this chapter to resistive circuits. These circuits will be powered by constant (dc) voltage and current sources.

1.2 Voltage

Consider two parallel metal plates at a and b, with a net positive charge on a, and an equal and opposite charge on b, as shown in Fig. 1.1. If U_{ab} is the work done, in joules, in moving a point charge of $+Q$ coulombs from b to a, against the force of repulsion between $+Q$ and the positive charge at a and the force of attraction between $+Q$ and the negative charge at b, then V_{ab} is the potential difference between points *a* and *b*. V_{ab} is also referred to as the voltage of *a* with respect to *b*, where

$$V_{ab} = \frac{U_{ab}}{Q} \qquad (1.1)$$

$$V_{ab} = -V_{ba} \qquad (1.2)$$

V_{ab} is measured in volts, abbreviated V. If *a* is more positive than *b*, then V_{ab} is represented with an arrow pointing to *a*, thus defining the polarity between the two points by specifying which point is at a higher electrical potential than the other.

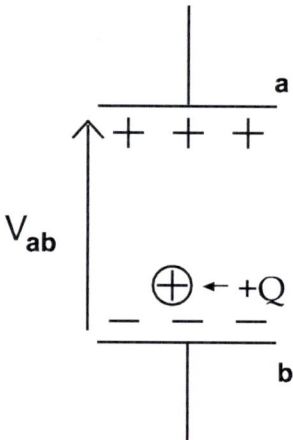

Fig.1.1 V_{ab}, the voltage of point *a* with respect to point *b*.

1.3 Ideal sources

A charged battery is a voltage source that can energize a circuit. It is represented by the symbol of Fig. 1.2. V_{ab} provided by the independent, ideal voltage source of Fig. 1.2 is always constant in magnitude and polarity (as given by the voltage arrow drawn next to it), whereas the magnitude and direction of the current *I*, that it provides, is determined by the circuit that the source is connected to. If *I* is in the direction shown, with the current flowing out of the positive terminal, then the source is a generator of electrical energy. If *I* flows into the positive terminal, the source is acting as an absorber of energy. A deactivated voltage source is represented by a short circuit, when used in the context of the superposition, Thévenin or Norton theorems that will be discussed later.

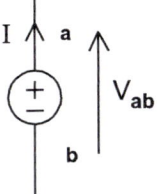

Fig. 1.2 Ideal voltage source

Electrical energy can also be provided to a circuit by a current source. The magnitude of the current *I*, supplied by the independent, ideal current source, shown in Fig.1.3, is always constant, and the direction of this current is always that indicated by the arrow within the symbol. The magnitude and polarity, however, of the voltage V_{ab} appearing across the terminals of the source, are determined by circuit conditions.

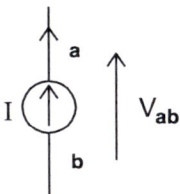

Fig. 1.3 Ideal current source

V_{ab}, as drawn in Fig.1.3, assumes that terminal *a* of the current source is at a higher potential than *b*. Whether this assumption is true or false is determined by the circuit that the source is connected to. If the circuit analysis finds V_{ab} to be a positive number, then the assumption is true, and *a* is at a higher potential than *b*. It follows from this that since the current flows out of the positive terminal of the source, then the source is acting as a generator. If the analysis finds V_{ab} to be negative, then it means that the assumption about the direction of the voltage arrow is wrong, and that terminal *b* is more positive than *a*. Under these conditions the current flows into the positive terminal of the current source, and the source acts as an absorber of electrical energy.

A deactivated current source is represented in a circuit by an open circuit, that is by two terminals that do not touch one another.

1.4 Ohm's law, the resistor and electrical power

The rate of flow of positive charge, in coulombs per second, through the cross-sectional area of a wire, constitutes the current flowing in the wire. The direction of the current is in the same direction as the flow of the positive charge. The resistance to the flow of charge varies between materials. Copper, gold and silver offer very low resistance to the flow of current and are considered conductors. An alloy of nickel, chromium, iron and carbon (Nichrome) offers higher resistance to the current and is called a resistor. Therefore, a length of copper wire connecting two points in an electric circuit is drawn as a straight line in a circuit diagram, and represents a connection of zero resistance to the flow of current. A resistor is represented by the symbol of Fig. 1.4 and has a resistance of *R* ohms (abbreviated Ω).

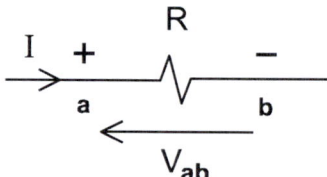

Fig. 1.4 A schematic representation of a resistor, of resistance *R* ohms, where a current of *I* amperes flows through the resistor and a voltage of V_{ab} volts appears across the resistor.

R is a measure of the resistance to the flow of current through the resistor. For a fixed voltage applied across a resistor, the higher the resistance, the lower the current passing through the resistor. The resistor is a linear element in that the current passing through it is directly proportional to the voltage applied across it.

The current enters the resistor at terminal a and leaves at b. Therefore, a is at a higher potential than b. The voltage arrow V_{ab} is then drawn to point to the terminal of entry of the current into the resistor. The voltage arrow defines the polarity associated with the current-carrying resistor; by pointing at a terminal it indicates that that terminal is at a higher electrical potential, or is more positive, than the other terminal of the element. The voltage appearing across the resistor is also referred to as the voltage drop, or the voltage dropped, across the resistor.

The relationship between the voltage appearing across the resistor and the current flowing through it, is given by Ohm's law

$$V_{ab} = IR \tag{1.3}$$

where the current is in amperes, abbreviated A.

The resistor is a passive circuit element. It always absorbs electrical energy. This energy is irreversibly converted to heat. The rate of absorption of energy, in joules per second, is the power P, in watts (W), where

$$P = V_{ab}I \tag{1.4}$$

In general, for constant or dc (direct current) currents and voltages, Ohm's law and Equ.1.4 give

$$P = VI = I^2R = \frac{V^2}{R} \tag{1.5}$$

1.5 Kirchhoff's laws

Kirchhoff's current law (KCL) states that the algebraic sum of currents at a node is zero. Therefore, by KCL the current entering node a, in Fig. 1.5, equals the sum of the two branch currents leaving it.

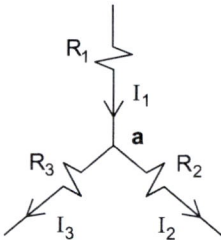

Fig.1.5 KCL gives $I_1=I_2+I_3$

Kirchhoff's voltage law (KVL) states that the algebraic sum of the voltages around a loop is zero.

Example 1.1: KCL, KVL, Ohm's law

Apply KVL to loop $abcd$, KCL at node b, and find an expression for the voltage V appearing across the current source, in the circuit of Fig. 1.6.

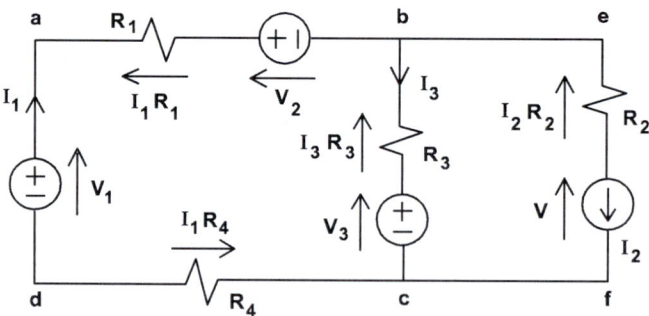

Fig. 1.6 Circuit for Example 1.1.

KVL applied to loop *abcd* gives

$$V_1 - I_1 R_1 - V_2 - I_3 R_3 - V_3 - I_1 R_4 = 0 \qquad (1.6)$$

KCL at node *b* gives

$$I_3 = I_1 - I_2 \qquad (1.7)$$

The voltage V appearing across the current source is found by applying KVL to loop *befc* so that

$$V = V_3 + I_3 R_3 - I_2 R_2 \qquad (1.8)$$

If V, as shown in Fig. 1.6 is calculated to be positive, then the assumption about the direction of the voltage arrow is correct and the source is then acting as an absorber. However, if the voltage arrow is found to be pointing in the opposite direction to that assumed in the circuit, then the source is a generator. If the direction of I_1 is as assumed in the circuit, then V_1 is a generator while V_2 is an absorber of energy.

Example 1.2: KVL

Express V_{ab} in terms of the voltages given in Fig. 1.7.

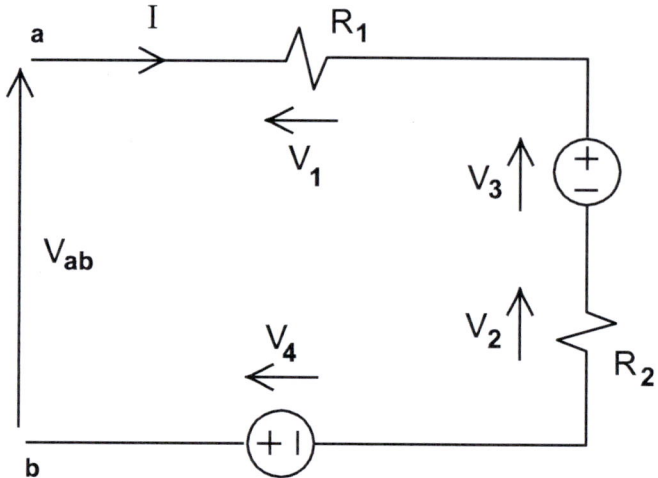

Fig. 1.7 Using KVL to find V_{ab} (Example 1.2).

KVL can be used to relate the voltage difference between any two points in a loop to the algebraic sum of voltage drops across individual circuit element between the two points, so that V_{ab} in Fig. 1.7 is given by:

$$V_{ab} = V_1 + V_2 + V_3 - V_4 \qquad (1.9)$$

1.6 Analysis of single-loop and single-node-pair circuits

KCL, KVL and Ohm's law can be used to determine the currents and voltages associated with elements arranged in a single loop or connected between two nodes.

Example 1.3: Single-loop method

Find an expression for the current I_1 in the circuit of Fig. 1.8. Discuss the circuit.

Fig. 1.8 shows a single-loop circuit.

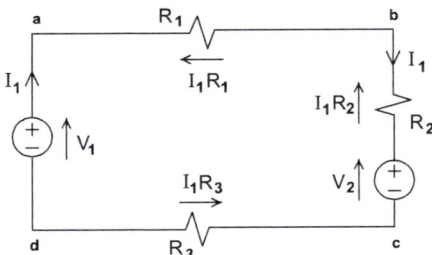

Fig. 1.8   A single-loop circuit with two opposing voltage sources and resistors in series (Example 1.3).

The current flowing through the loop is assumed to be in the clockwise direction. The voltage drops across the resistors are then in the directions shown and of magnitudes indicated according to Ohm's law. KVL applied to the loop then gives:

$$V_1 - I_1 R_1 - I_1 R_2 - V_2 - I_1 R_3 = 0 \qquad (1.10)$$

$$I_1 = \frac{V_1 - V_2}{R_1 + R_2 + R_3} \qquad (1.11)$$

If the numerical calculation results in a positive value for the current, then the assumption about its direction, and the resulting directions for the voltage arrows across resistors, is as assumed in Fig. 1.8. A negative current means that the assumption about the direction of the current is wrong, and that the voltage arrows, showing voltage drops across the resistors, have to be reversed. Equ. 1.11 is equivalent to Ohm's law. The current through each of the three resistors is the same and the resistors are said to be in series. The resistors can then be replaced with a single equivalent resistor R_{eq} whose value is the sum of the three resistors. The difference between V_1 and V_2 is the total driving force that causes the current to flow through R_{eq}. If $V_1 > V_2$, then the current will flow in a clockwise direction. If $V_1 < V_2$, an anticlockwise current will result.

Power is conserved in a circuit so that the power generated is equal to the power consumed. Assuming $V_1 > V_2$, in the circuit of Fig. 1.8, then the clockwise current that flows in the circuit means that V_1 is a generator, and V_2 an absorber, of electrical power; the former generating $V_1 I_1$ watts and the latter absorbing $V_2 I_1$ watts. The energy balance equation then gives:

$$V_1 I_1 = V_2 I_1 + I_1^2 R_1 + I_1^2 R_2 + I_1^2 R_3 \qquad (1.12)$$

Example 1.4: Single-node-pair method

Find an expression for Vab, in Fig. 1.9, in terms of I, R_1 and R_2. Discuss the circuit in relation to Ohm's law.

The circuit of Fig. 1.9 can be analyzed using the single-node-pair method. The circuit has two nodes; one at a and another at b. V_{ab} is the voltage of node a with respect to node b. V_{ab} appears across the terminals of each of resistors R_1 and R_2 and across the current source.

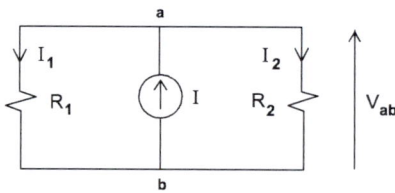

Fig. 1.9 A single-node-pair circuit (Example 1.4).

10

By KCL at node *a*

$$I = I_1 + I_2 \qquad (1.13)$$

Using Ohm's law for each resistor and substituting for I_1 and I_2 in Equ. 1.13 gives:

$$I = \frac{V_{ab}}{R_1} + \frac{V_{ab}}{R_2} \qquad (1.14)$$

Given I, R_1 and R_2, V_{ab}, I_1 and I_2 are calculated from:

$$V_{ab} = I \left(\frac{1}{(1/R_1) + (1/R_2)} \right) \qquad (1.15)$$

$$I_1 = V_{ab}/R_1 \qquad (1.16)$$

$$I_2 = V_{ab}/R_2 \qquad (1.17)$$

The same voltage appears across R_1 and across R_2. The current I, entering node *a*, divides into two currents that flow through R_1 and R_2. The two resistors are said to be connected in parallel. Resistors in parallel can be combined into an equivalent resistor R_{eq} so that for the two resistors in Fig. 1.9

$$R_{eq} = \frac{1}{((1/R_1) + (1/R_2))} \qquad (1.18)$$

Equation 1.15 can then be written as:

$$V_{ab} = IR_{eq} \qquad (1.19)$$

This represents Ohm's law.

Example 1.5: Single-node-pair method

Express V_{ab}, in Fig.1.10, in terms of the source currents and the resistances of the resistors. Discuss whether the sources are generating or absorbing electrical power.

Figure 1.10 shows a single-node-pair circuit involving two current sources.

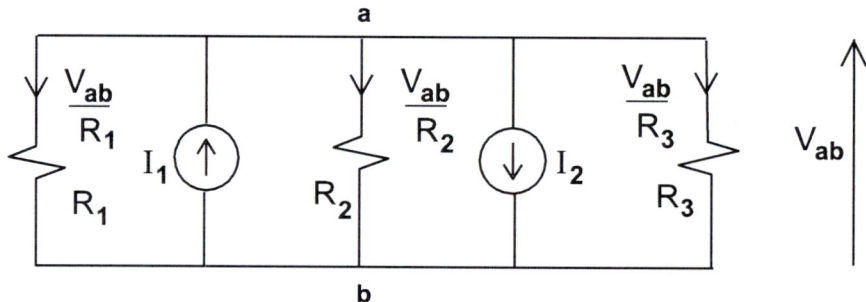

Fig 1.10 Two-source single-node-pair circuit (Example 1.5).

The upper terminals of the five elements of the circuit are all connected together and are, therefore, at the same electrical potential. The lower terminals of each element are also connected and form a separate node b. V_{ab} is then the voltage of node a with respect to node b. V_{ab} appears across each element in the circuit. Assuming that $I_1 > I_2$ then there is a net flow of current, (I_1-I_2), into node a. Equating the net flow of current into node a to the sum of currents flowing out through the resistors, using KCL, gives:

$$I_1 - I_2 = \frac{V_{ab}}{R_1} + \frac{V_{ab}}{R_2} + \frac{V_{ab}}{R_3}$$

$$V_{ab} = \frac{(I_1 - I_2)}{\left(\dfrac{1}{R_1} + \dfrac{1}{R_2} + \dfrac{1}{R_3} \right)} \qquad (1.20)$$

Given the values of the source currents and of the resistances of the resistors, V_{ab} can be calculated. The same voltage appears across each resistor and so the resistors are connected in parallel. Equation 1.20 is a statement of Ohm's law. It relates the net current flow into the node to the voltage difference across the equivalence of the three resistors in parallel. If $I_1 > I_2$ then V_{ab}, as shown in Fig. 1.10, is positive. If $I_1 < I_2$ then V_{ab} is negative and the voltage arrow should be shown pointing at node b to indicate that V_{ba} is positive. In the latter case the net flow of current is into node b and then up through the three resistors.

The concept of a representing a single node can be visualized by shrinking the horizontal line at a, to which the upper terminals of the five elements are connected to in Fig. 1.10, to a point. The concept of one current flowing in and four currents flowing out of a single point can then be seen. Stretching this point back to the horizontal conductor of Fig. 1.10 creates additional nodes. All these nodes are, however, at the same potential, V_{ab}. KCL applies to all the new nodes and its application results in three equations, which now involve the additional currents between the new nodes. The elimination of these additional currents once again gives Equ. (1.20).

If $I_1 > I_2$ in the circuit of Fig. 1.10, then source I_1 is a generator and I_2 an absorber of energy. The energy balance equation then gives:

$$V_{ab}I_1 = V_{ab}I_2 + V_{ab}^{\;2}\left(\frac{1}{R_1} + \frac{1}{R_2} + \frac{1}{R_3} \right) \qquad (1.21)$$

13

1.7 Series and parallel combination of resistors
1.7.1 Series combination

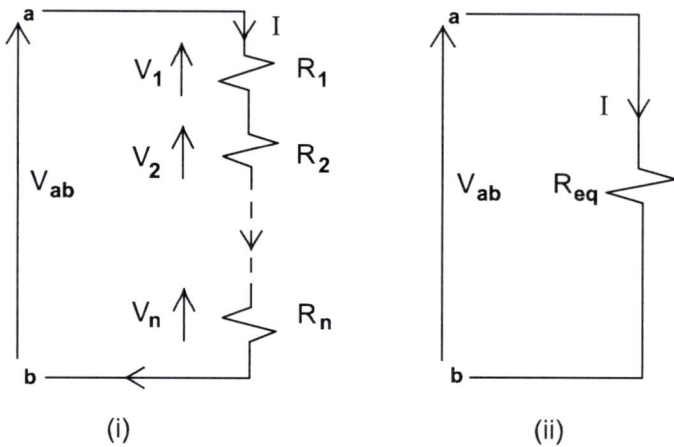

Fig.1.11 The series resistors in (i) can be replaced with the equivalent resistor R_{eq} in (ii).

The same current flows through the resistors connected between terminals *a* and *b*, in Fig. 1.11(i), and the resistors are said to be in series with each other. The resistors in series can be replaced with an equivalent resistor R_{eq}, so the circuit between *a* and *b* is represented by Fig. 1.11(ii). In Fig. 1.11(i)

$$V_{ab} = V_1 + V_2 + ... + V_n \qquad (1.22)$$

Using Ohm's law

$$V_{ab} = IR_1 + IR_2 + ... + IR_n \qquad (1.23)$$
$$V_{ab} = I(R_1 + R_2 + ... + R_n) \qquad (1.24)$$
$$V_{ab} = IR_{eq} \qquad (1.25)$$

Therefore,

14

$$R_{eq} = R_1 + R_2 + ... + R_n \qquad (1.26)$$

Equation 1.26 shows that the equivalent resistor, for resistors in series, is equal to the sum of the individual resistors.

1.7.2 Parallel combination

The resistors in the circuit of Fig. 1.12(i) are connected in parallel. The voltage across each resistor is the same, and the current flowing into node a divides into the currents flowing through the individual resistors. The resistors to the right of terminals a and b, in Fig. 1.12(i), can be combined into an equivalent resistor, R_{eq}, that is shown in Fig. 1.12(ii).

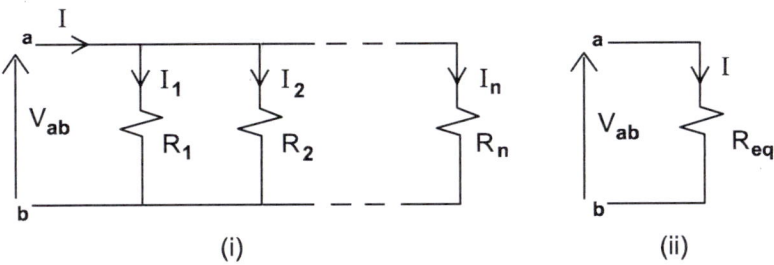

(i)   (ii)

Fig. 1.12 Resistors in parallel in (i) replaced by their equivalent resistor in (ii).

KCL in Fig. 1.12(i) gives:

$$I = I_1 + I_2 + ... + I_n \qquad (1.27)$$

$$I = V_{ab}\left(\frac{1}{R_1} + \frac{1}{R_2} + ... + \frac{1}{R_n}\right) \qquad (1.28)$$

$$V_{ab} = I\left(\frac{1}{1/R_1 + 1/R_2 + ... + 1/R_n}\right) \qquad (1.29)$$

$$V_{ab} = IR_{eq} \qquad (1.30)$$

15

Therefore, the equivalent resistance is given by:

$$R_{eq} = \cfrac{1}{\left(\cfrac{1}{R_1} + \cfrac{1}{R_2} + ... \cfrac{1}{R_n}\right)} \qquad (1.31)$$

Equation 1.31 is in a form that is easy to use in both dc and in phasor calculations, where the latter is used in the analysis of ac (alternating current) circuits.

The equivalent resistance of two equal value resistors that are connected in parallel is equal to half the value of one of the resistors. The parallel equivalent resistance of two different value resistors is less than the smaller of the two resistors. The parallel equivalent resistance of a resistor with a short circuit is zero.

Example 1.6: Equivalent resistance

Find the expression of the equivalent resistance to the left of terminals a and b, of the circuit of Fig. 1.13, in terms of the resistors given. Discuss the usefulness of this expression.

If + stands for a series and || stands for a parallel combination of resistors, then, R_{eq} is given by:

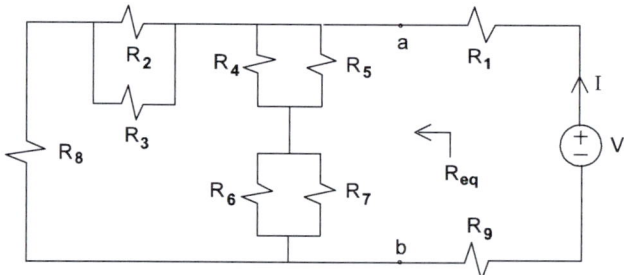

Fig. 1.13 Replacing the circuit to the left of terminals a and b with an equivalent resistance, R_{eq}, reduces the circuit to a single-loop (Example 1.6).

16

$$R_{eq} = \{[R_8 + (R_2 \| R_3)] \| [(R_4 \| R_5) + (R_6 \| R_7)]\} \qquad (1.32)$$

$$R_{eq} = 1/\left\{\left[\frac{1}{R_8 + 1/(1/R_2 + 1/R_3)}\right] + \left[\frac{1}{1/(1/R_4 + 1/R_5) + 1/(1/R_6 + 1/R_7)}\right]\right\}$$

(1.33)

  This method of writing the expression of the equivalent resistance is ideal for calculations because of its ease of use in modern calculators.

  If the aim were to find the current I, in the circuit of Fig. 1.13, then reducing the circuit to the left of a and b to an equivalent resistance, and then reconnecting this resistance between a and b, gives a single-loop circuit where I can be found using the single-loop method.

Example 1.7: Circuit analysis using equivalent resistance (circuit reduction method)

Combine resistors to reduce the right hand side of the circuit of Fig. 1.14(i) to a single equivalent resistor, to transform the circuit to a single-loop, to find the current I. Then find I_1, V_{bc} and V_{ac}.

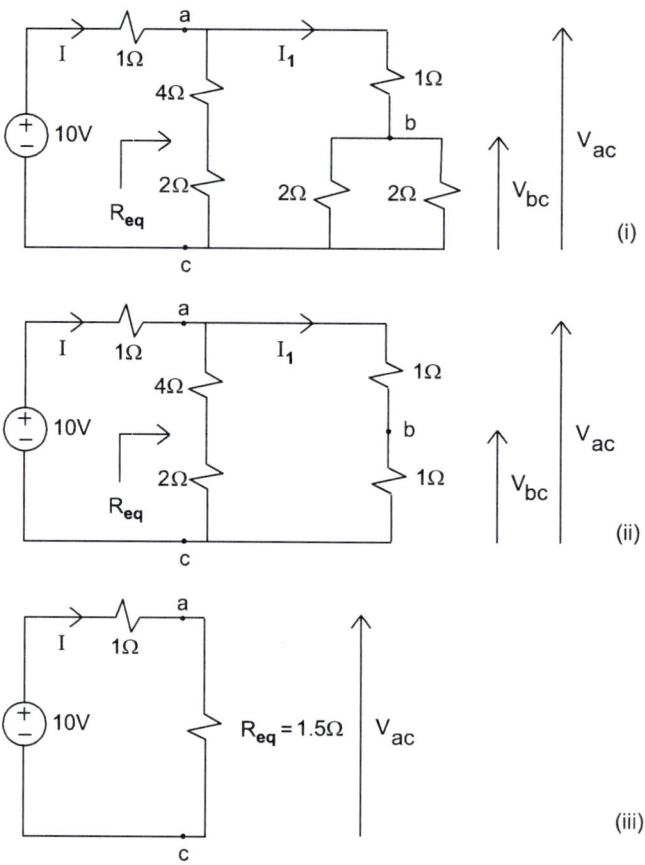

Fig. 1.14 Reduction of circuit in (i) to a single loop in (iii) (Example 1.7).

18

Combining the two 2Ω resistors of Fig. 1.14(i) in parallel results in the circuit of Fig. 1.14(ii). Then combining the 6Ω and 2Ω resistors, between terminals a and b, in parallel (in Fig. 1.14(ii)), gives the equivalent resistance of the circuit to the right of terminals a and b to be

$$R_{eq} = \left(\frac{1}{(1/6)+(1/2)} \right) \qquad (1.34)$$

$$R_{eq} = 1.5\Omega \qquad (1.35)$$

The circuit then reduces to a single loop (Fig. 1.14(iii)) and Ohm's law gives

$$I = 10/(1+1.5) \qquad (1.36)$$

$$I = 4\,\text{A} \qquad (1.37)$$

$$V_{ac} = IR_{eq} = 4 \times 1.5 \qquad (1.38)$$

$$V_{ac} = 6\,\text{V} \qquad (1.39)$$

Returning to the circuit of Fig. 1.14(ii)

$$I_1 = V_{ac}/2 = 6/2 \qquad (1.40)$$

$$I_1 = 3\,\text{A} \qquad (1.41)$$

$$V_{bc} = I_1 \times 1 = 3 \times 1 \qquad (1.42)$$

$$V_{bc} = 3\ \text{V} \qquad (1.43)$$

1.8 Voltage and current division

The source voltage V_{ab} is dropped across the resistors R_1 to R_n connected in series in the circuit of Fig 1.15. Since the current through all resistors is the same

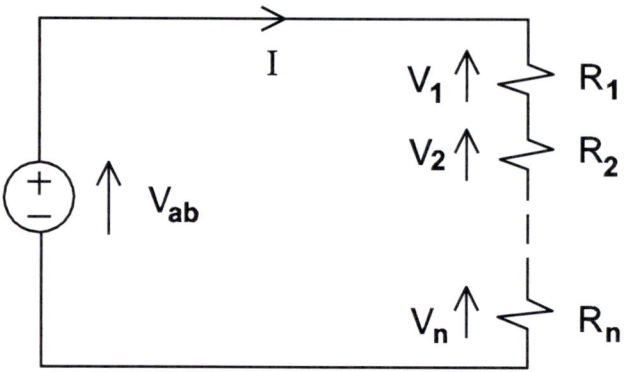

Fig 1.15 V_{ab} is dropped across resistors connected in series.

$$V_{ab} = I(R_1 + R_2 + ... + R_n) \qquad (1.44)$$
$$V_x = IR_x \qquad (1.45)$$

where V_x is the voltage dropped across a general resistor R_x. Dividing Equ. 1.45 by Equ 1.44 gives

$$V_x = \frac{R_x}{(R_1 + R_2 + ... + R_n)} V_{ab} \qquad (1.46)$$

Equation 1.46 is the voltage division equation. It gives the voltage dropped across an individual resistor in a combination of resistors in series, provided that the voltage across the combination is known.

20

The current entering node a in Fig. 1.16 divides into individual currents that flow through the resistors R_1 to R_n that are connected in parallel. The voltage across each resistor is V. Applying KCL at a and Ohm's law for each resistor gives:

$$I = V\left(\frac{1}{R_1} + \frac{1}{R_2} + \dots \frac{1}{R_n}\right)$$

(1.47)

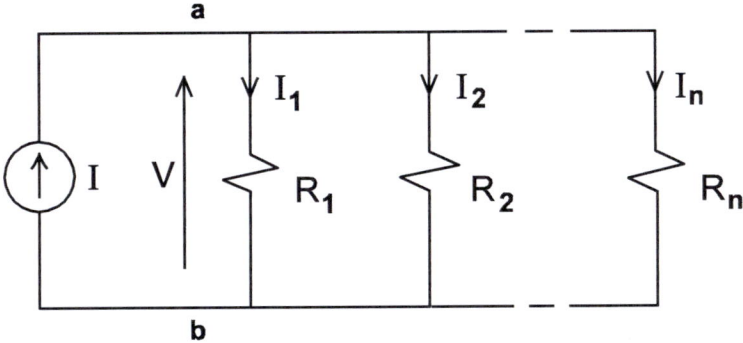

Fig.1.16 The source current I divides into resistor currents.

If the current through a general resistor R_x is I_x, then Ohm's law gives:

$$I_x = \frac{V}{R_x}$$

(1.48)

Dividing Equ. 1.48 by Equ. 1.47 gives

$$I_x = \left(\frac{1/R_x}{1/R_1 + 1/R_2 + \dots + 1/R_n}\right)I$$

(1.49)

Equation 1.49 is the current division equation.

1.9 Superposition theorem

The superposition theorem can be applied to the analysis of linear circuits and systems. Consider a system where, if the input is f_{in1} then the output is f_{out1}, and if the input is f_{in2} the output is f_{out2}. For the system to be linear, if the input is now Af_{in1}, where A is a constant, then the corresponding output should be Af_{out1}, and if the input is $(f_{in1} + f_{in2})$ then the corresponding output should be $(f_{out1} + f_{out2})$. The circuit components and circuits considered in this work are linear.

Superposition is a powerful method for the analysis of electrical circuits which contain multiple energy sources. It is used to find the current and voltage associated with a particular element in the circuit. It assumes that each source contributes a component to, for instance, the current in a branch of the circuit. A current component is found by analyzing the circuit with a single source active, and all other sources deactivated. The other sources are then considered, one at a time, to find their contributions to the current. The sum of individual current components gives the current when all sources are active simultaneously. Deactivation of sources involves replacing current sources by open circuits and voltage sources by short circuits.

Example 1.8: Superposition

Use superposition in the multiple source circuit of Fig. 1.17 to find the current I_a flowing through R_3, and the voltage V_{ab} appearing across R_3.

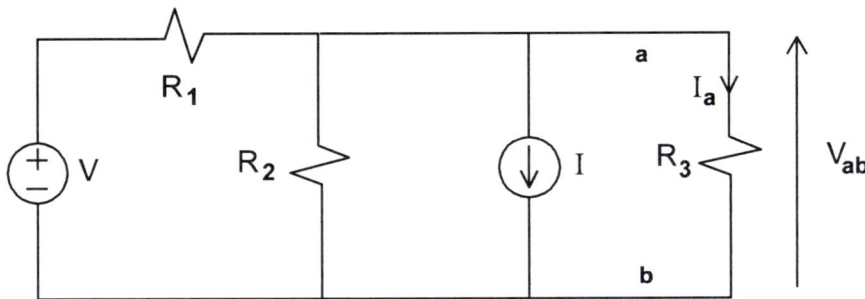

Fig. 1.17 A two-source circuit where superposition can be used as a circuit analysis method (Example 1.8).

To find the component of current I_{a1} and of voltage V_{ab1}, associated with R_3, due to the voltage source V acting alone, the current source in Fig. 1.17 is deactivated resulting in the circuit of Fig. 1.18.

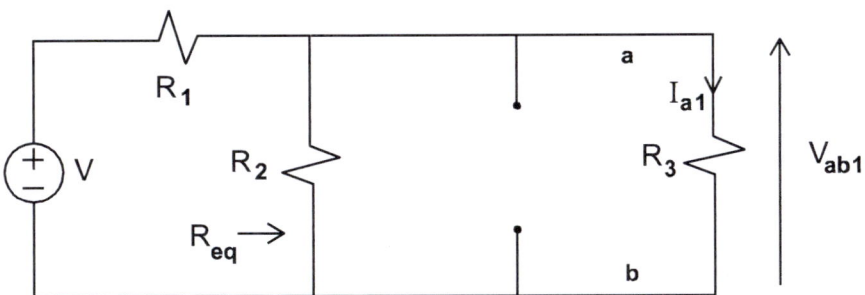

Fig. 1.18 The circuit of Fig. 1.17 with the current source deactivated (Example 8).

If R_{eq} is the parallel equivalent resistance of R_2 and R_3, then,

$$R_{eq} = \frac{1}{\left(\dfrac{1}{R_2} + \dfrac{1}{R_3}\right)}$$ (1.50)

and voltage division gives

$$V_{ab1} = \left\{ \frac{\left(\dfrac{1}{1/R_2 + 1/R_3}\right)}{\left[\dfrac{1}{(1/R_2 + 1/R_3)} + R_1\right]} \right\} V$$ (1.51)

Since the voltage across R_3 is V_{ab1}, the current I_{a1} is

$$I_{a1} = \frac{V_{ab1}}{R_3}$$ (1.52)

To find the component of current and of voltage I_{a2} and V_{a2}, respectively, due to the current source I acting alone, the voltage source V in Fig. 1.17 is deactivated, resulting in the circuit of Fig. 1.19.

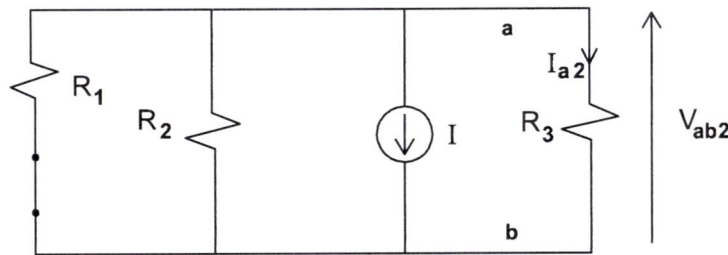

Fig. 1.19 The circuit of Fig. 1.17 with the voltage source deactivated (Example 8).

Current division then gives:

$$I_{a2} = -\left[\frac{(1/R_3)}{(1/R_1 + 1/R_2 + 1/R_3)}\right]I \qquad (1.53)$$

and Ohm's law gives the voltage component as

$$V_{ab2} = -R_3 I_{a2} \qquad (1.54)$$

The negative current and voltage in Equs. 1.53 and 1.54 imply that the arrows for I_{a2} and V_{ab2} are in fact pointing in the opposite direction to that assumed in Fig 1.19.

The total current and voltage, I_a and V_{ab}, in the circuit of Fig. 1.17, are then obtained by substituting for current and voltage components, given in Equs. 1.51 to 1.54, with their signs as appearing in these equations, into Equs. 1.55 and 1.56, where

$$I_a = I_{a1} + I_{a2} \qquad (1.55)$$
$$V_{ab} = V_{ab1} + V_{ab2} \qquad (1.56)$$

25

Example 1.9: Superposition

Find the current I in the circuit of Fig. 1.20(i).

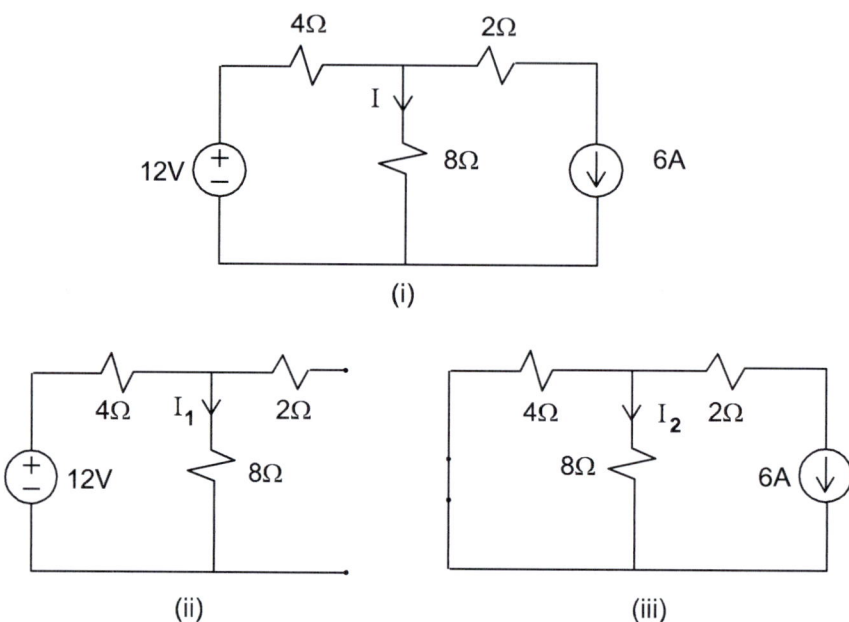

Fig. 1.20. Example 1.9.

Let I_1 be the component of I due to the 12V source acting alone and I_2 the component due to the 6A source acting alone. From superposition $I=I_1+I_2$. From Fig. 1.2(ii), Ohm's law gives

$$I_1 = 12/(4+8) = 1A$$

From the circuit of Fig. 1.20(iii) current division gives

$$I_2 = -\{(1/8)/[(1/8)+(1/4)]\}6 = -2A$$
$$I = 1-2 = -1A$$

1.10 Thévenin's and Norton's theorems

Thévenin's theorem states that any two-terminal linear circuit, say contained within the box in Fig.1.21(i), can be replaced with an equivalent circuit that comprises a single independent voltage source in series with a single resistance as in Fig.1.21(ii).

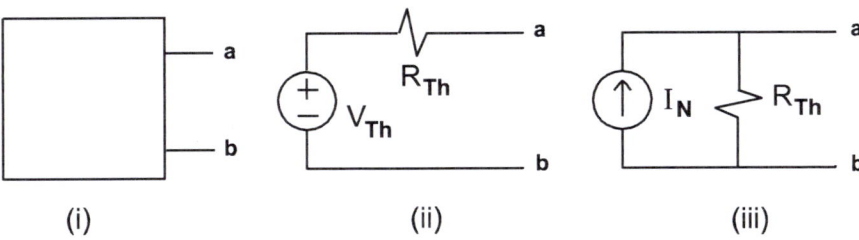

(i)   (ii)   (iii)

Fig. 1.21 The two-terminal linear network of (i) replaced by its Thévenin equivalent circuit in (ii) and its Norton equivalent in (iii).

If a resistor, or load, is connected between the two terminals, the equivalent circuit produces the same effect in the resistor as the original circuit. The equivalent circuit is called the Thévenin equivalent of the circuit to the left of terminals a and b in Fig. 1.21, or the Thévenin equivalent of the circuit looking into the two terminals of the circuit. The voltage and resistance of the Thévenin equivalent circuit are called the Thévenin equivalent voltage V_{Th}, and the Thévenin equivalent resistance R_{Th}, respectively [Fig. 1.21(ii)].

Norton's theorem states that the two-terminal circuit can be replaced with an independent current source I_N, in parallel with the Thévenin resistance [Fig. 1.21(iii)].

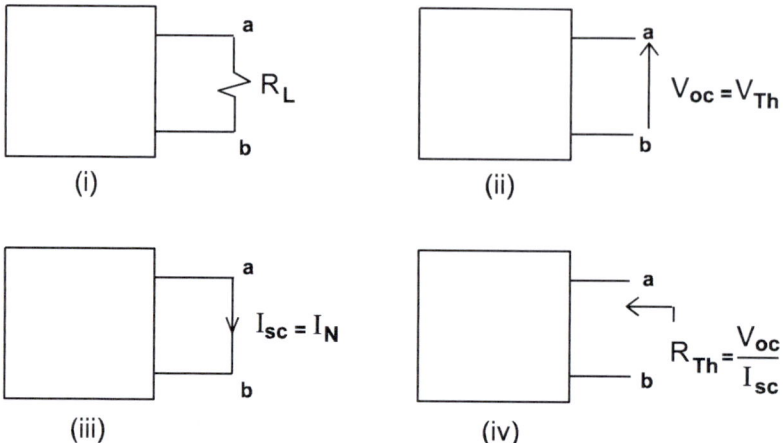

Fig. 1.22 Method of determining the parameters of the Thévenin and Norton equivalents for the circuit seen by the load resistance R_L.

Fig. 1.22(i) shows a load resistor R_L driven by a two-terminal circuit. The Thévenin and Norton equivalent circuits of the network to the left of terminals a and b are determined by first removing R_L. The Thévenin equivalent voltage is then determined by keeping the terminals a and b open and by measuring the open circuit voltage that appears between a and b. The open circuit voltage V_{oc} is the Thévenin equivalent voltage V_{Th} [Fig. 1.22(ii)]. The short-circuit current I_{sc}, that flows between a and b, is the Norton equivalent current I_N [Fig. 1.22(iii)]. The equivalent resistance of the circuit looking into terminals a and b, with a and b on open circuit, and with sources in the circuit deactivated (voltage sources replaced by short circuits and current sources replaced by open circuits), is the Thévenin equivalent resistance [Fig. 1.22(iv)]. R_{Th} is also equal to (V_{oc}/I_{sc}). Reconnecting R_L across terminals a and b, of either of the resulting Thévenin or Norton equivalent circuits, gives the current, voltage or power associated with R_L, that would have resulted had the original system been driving R_L.

Example 1.10: Thévenin equivalent

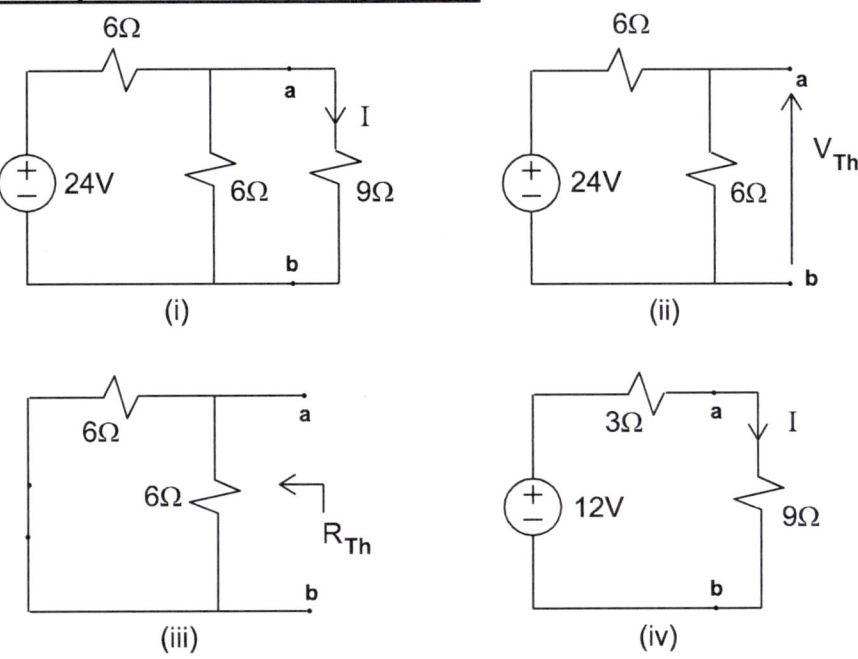

Fig. 1.23 Find I in (i). (Example 1.10).

The method to be used is to replace the circuit seen by the 9Ω resistor with its Thévenin equivalent circuit, and to then reconnect the 9Ω resistor to this circuit to get a single loop. The current in this loop would be I. Removing the 9Ω resistor in the circuit of Fig. 1.23(i), and keeping terminals a and b on open circuit, the Thévenin voltage of the circuit to the left of terminals a and b is given, by voltage division in Fig. 1.23(ii), as $V_{Th} = \{6/(6+6)\}24 = 12\,\text{V}$. R_{Th} is the equivalent resistance of the circuit to the left of terminals a and b with the voltage source deactivated. R_{Th} is the equivalent of the two 6Ω resistors in parallel i.e. 3Ω [Fig. 1.23(iii)]. Reconnecting the 9Ω resistor to the Thévenin equivalent circuit [Fig. 1.23(iv)], and applying Ohm's law gives $I = [12/(9+3)] = 1\text{A}$.

Example 1.11: Thévenin equivalent

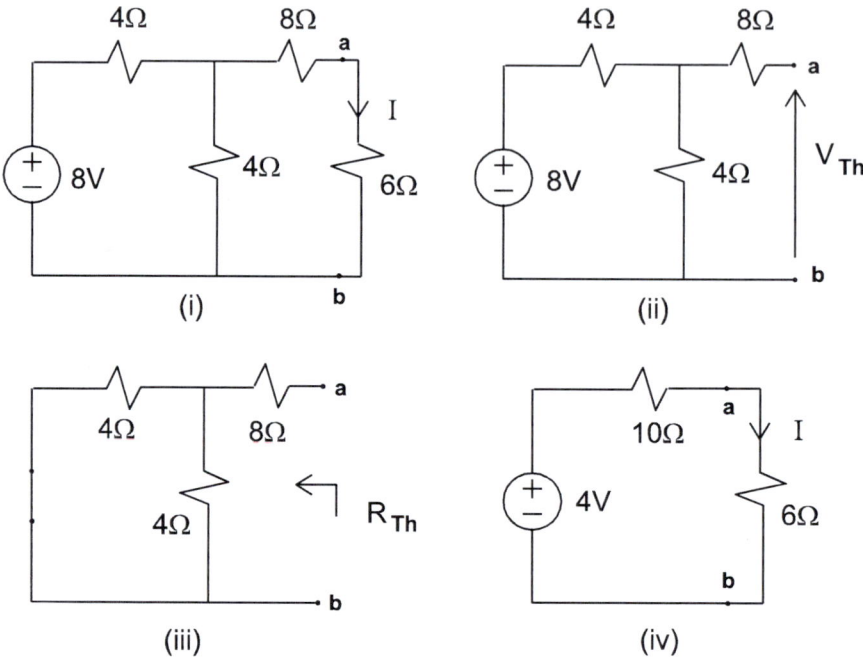

Fig. 1.24 Find I in (i). (Example 1.11).

   Removing the 6Ω resistor of Fig. 1.24(i), to replace the circuit to the left of terminals a and b with its Thévenin equivalent, voltage division gives the open circuit voltage between a and b, in Fig. 1.24(ii), as V_{Th}= [4/ (4+4)] 8=4V. Deactivating the 8V source and finding the equivalent resistance of the circuit to the left of terminals a and b, in Fig. 1.24(iii), gives R_{Th}=8+4‖4=8+2=10Ω. Reconnecting the 6Ω resistor to the Thévenin equivalent circuit, in Fig.1.24(iv), gives I= [4/ (6+10)] =0.25A.

Example 1.12: Thévenin equivalent

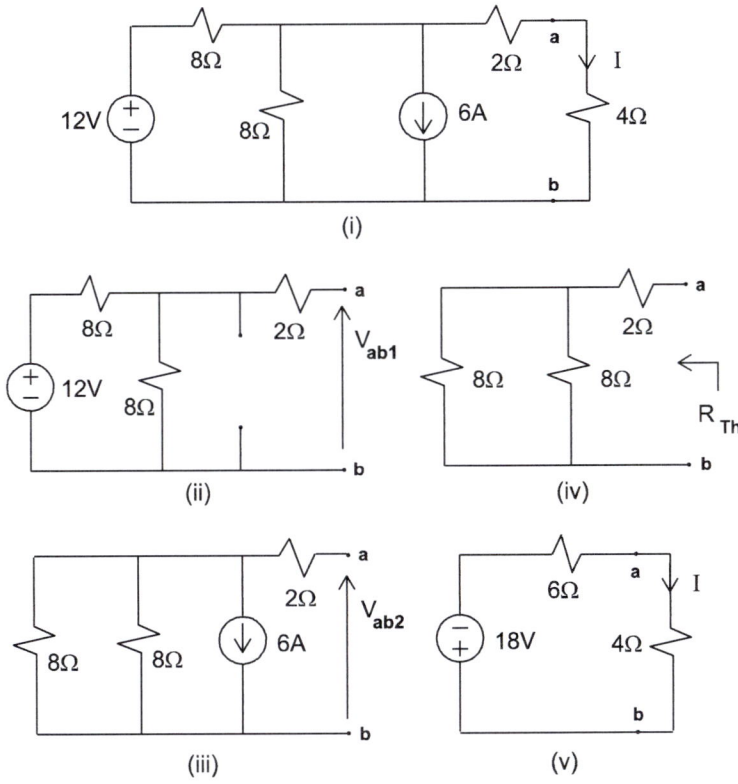

Fig.1.25. Find I in (i) (Example 1.12).

The Thévenin voltage of the circuit to the left of terminals a and b, with the 4Ω resistor of Fig. 1.25(i) removed, is, from superposition, $(V_{Th1} + V_{Th2})$. V_{Th1} and V_{Th2} are the open circuit voltages due to the 12V source and the 6A source acting alone, respectively. From Fig. 1.25(ii), $V_{Th1} = [8/(8+8)]12 = 6$ V, and from Fig. 1.25(iii), $V_{Th2} = -(8\|8)6 = -4 \times 6 = -24$ V. From Fig. 1.25(iv), $R_{Th} = 2 + (8\|8) = 2 + 4 = 6 \Omega$. Reconnecting the 4Ω resistor to the Thévenin equivalent circuit gives, $I = [(-24 + 6)/10] = -1.8$ A.

31

Example 1.13: Norton equivalent

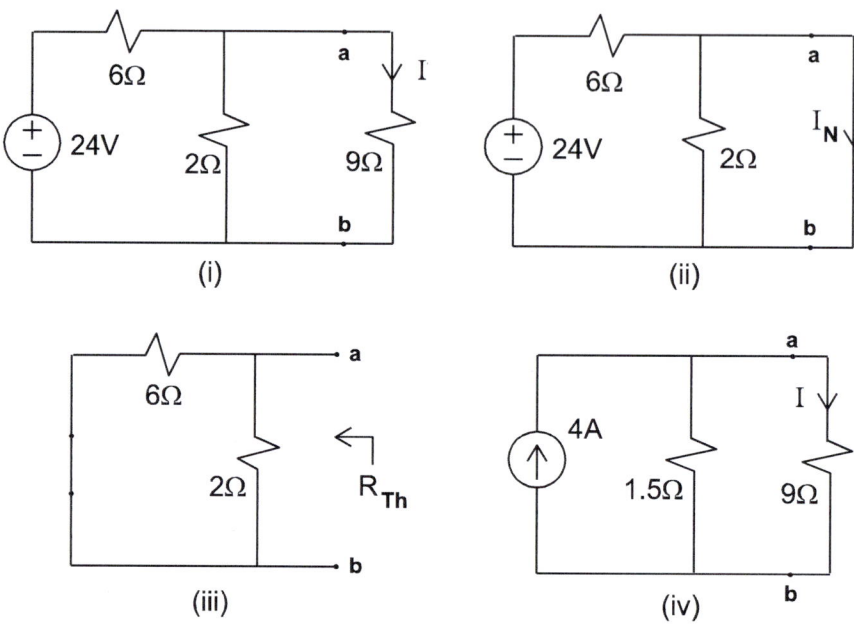

Fig. 1.26 Find *I* in (i) (Example 1.13).

Removing the 9Ω resistor of Fig. 1.26(i), to replace the circuit to the left of terminals *a* and *b* with its Norton equivalent, Fig. 1.26(ii) gives I_N, the short circuit current between *a* and *b* when these terminals are short circuited together, as $I_N = (24/6)= 4A$. Fig. 1.26(iii) gives $R_{Th}=6||2=\{1/[(1/6)+(1/2)]\}=1.5Ω$. Reconnecting the 9Ω resistor to the Norton equivalent circuit in Fig. 1.26(iv), and using current division gives $I= \{(1/9)/ [(1/9) + (1/1.5)]\} \times 4=0.57A$.

Example 1.14: Norton equivalent

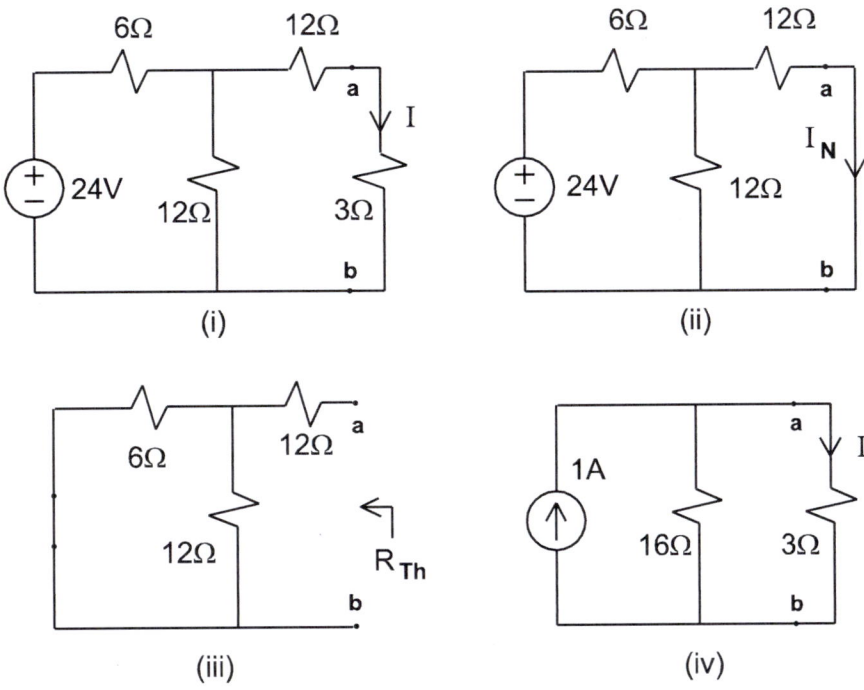

Fig. 1.27 Find I in (i) (Example 1.14).

Removing the 3Ω resistor of Fig. 1.27(i), to replace the circuit to the left of terminals a and b with its Norton equivalent circuit, Fig. 1.27(ii) gives I_N =0.5{24/ [6+ (12||12)]} = 0.5{24/ [6+6]} =1A. From Fig. 1.27(iii) R_{Th}=12+ (6||12) =12+4=16Ω. Reconnecting the 3Ω resistor to the Norton equivalent circuit, in Fig. 1.27(iv), gives I= {(1/3)/ [(1/16) + (1/3)]} ×1=0.8421A.

Example 1.15: Thévenin equivalent

Find the current *I*, in the circuit of Fig. 1.28, by using the Thévenin equivalent of the circuit to the left of terminals *a* and *b*.

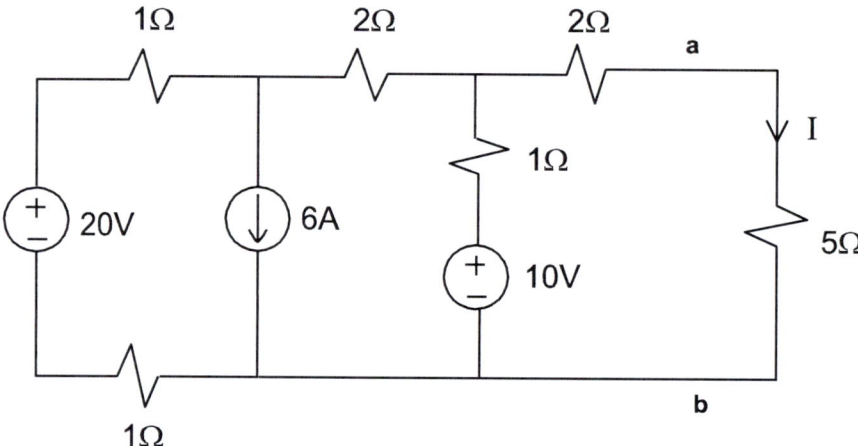

Fig. 1.28. The current *I* can be found by reducing the circuit to a single loop by replacing the circuit to the left of terminals *a* and *b* by its Thévenin equivalent. Since the circuit contains multiple sources, the Thévenin equivalent can be found using the superposition theorem (Example 1.15).

a) V_{Th}

The 5Ω resistor is first removed and the 6A current source is deactivated to find the component of the open circuit voltage V_{oc1}, appearing across terminals a and b, due to the voltage sources (see Fig. 1.29). In this circuit, considering the voltage sources together, rather than one at a time, gives the same result quicker.

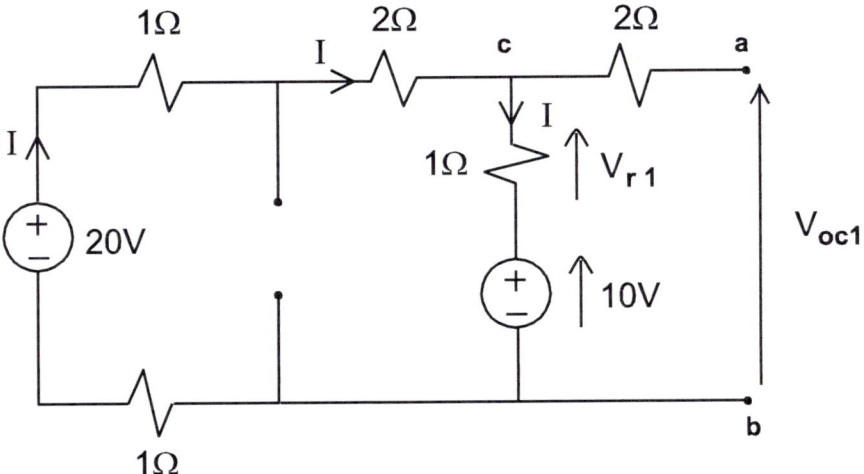

Fig. 1.29 Using superposition to find the component of the Thévenin voltage V_{oc1} due to the two voltage sources of Fig. 1.28 (Example 1.15).

Since no current flows in branch ca,

$$V_{oc1} = V_{ab} = V_{cb} = V_{Th1} \tag{1.57}$$

$$V_{oc1} = V_{r1} + 10 \tag{1.58}$$

$$V_{oc1} = \frac{(20-10)}{5} \times 1 + 10 \tag{1.59}$$

$$V_{oc1} = 12V \tag{1.60}$$

Considering now the effect of the current source on the Thévenin voltage, with the voltage sources in Fig. 1.28 deactivated, the circuit of Fig. 1.30 is obtained where current division, Ohm's law and KVL give

$$V_{oc2} = V_{Th2} = -V_{r2} = -\left\{ \frac{1/3}{[(1/3)+(1/2)]} \right\} \times 6 \times 1 \qquad (1.61)$$

$$V_{oc2} = -2.4 \text{ V} \qquad (1.62)$$

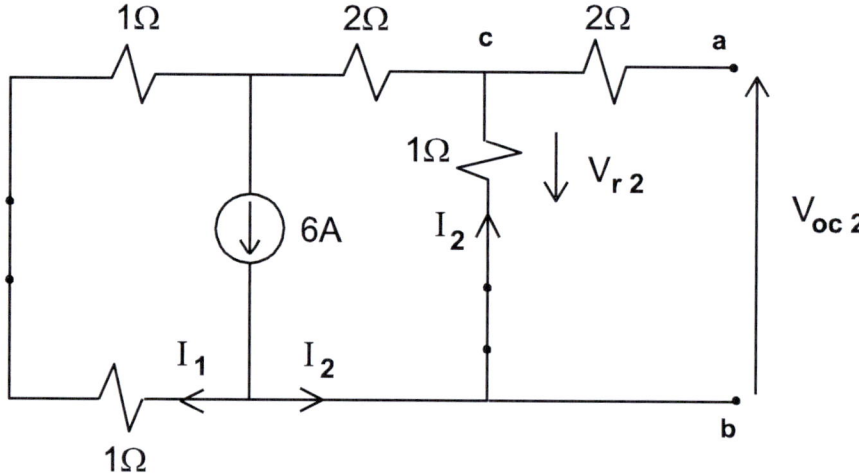

Fig. 1.30 Using superposition to find the component of the Thévenin voltage V_{oc2} due to the current source of Fig. 1.28 (Example 1.15).

The Thévenin voltage is then the sum of the two open circuit voltages V_{oc1} and V_{oc2} so that

$$V_{Th} = 12 - 2.4 = 9.6 \text{ V} \qquad (1.63)$$

b) R_Th

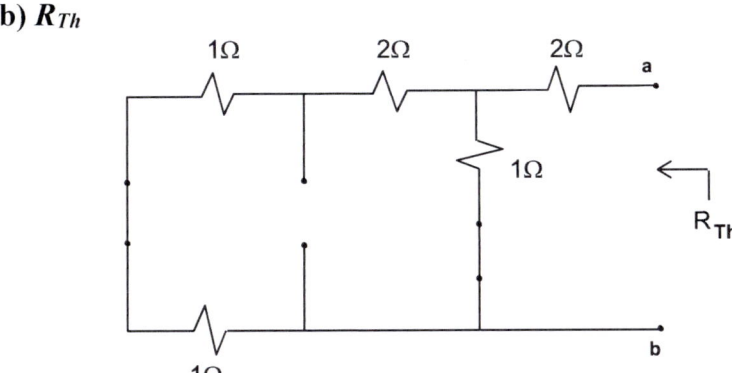

Fig. 1.31 Deactivating sources in circuit of Fig. 1.28 to find R_{Th} (Example 1.15).

R_{Th} is found by looking left into terminals a and b with R_L removed, and the sources in Fig. 1.28 deactivated (see Fig. 1.31) where

$$R_{Th} = 2 + \frac{1}{\left[1 + (1/4)\right]} \tag{1.64}$$

$$R_{Th} = 2.8\Omega \tag{1.65}$$

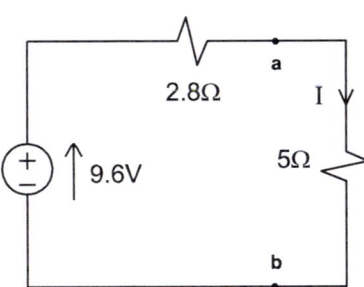

Fig. 1.32. The Thévenin equivalent circuit, of the circuit to the left of terminals a and b in Fig. 1.28, connected to the 5Ω load resistor (Example 1.15).

The single-loop circuit of Fig. 1.32 is obtained when the load resistor is connected to the Thévenin equivalent circuit. The current through the 5Ω resistor is then

$$I = \frac{9.6}{2.8 + 5} \tag{1.66}$$

$$I = 1.23 \text{ A} \tag{1.67}$$

Example 1.16: Thévenin and Norton equivalents

Find the Thévenin equivalent voltage V_{Th}, the Norton equivalent current I_N, and the Thévenin equivalent resistance R_{Th}, looking left into terminals a and b, in the circuit of Fig. 1.33.

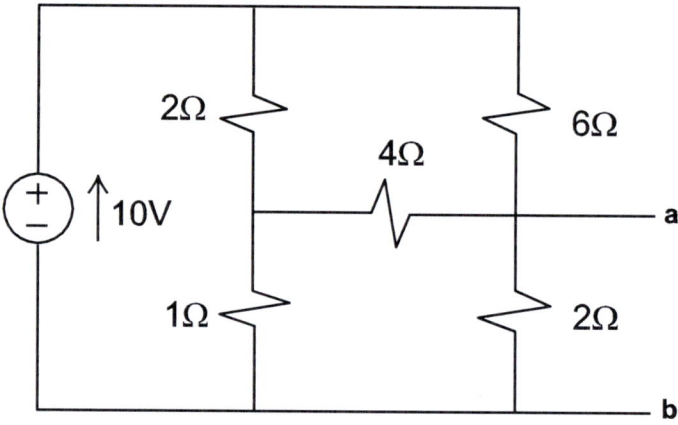

Fig. 1.33. Circuit for Example 1.16.

a) V_{Th}

The open-circuit voltage V_{oc}, for the circuit in Fig. 1.34, is V_{Th}. KVL for *dcb*, *dcab* and *dab* in Fig. 1.34 gives

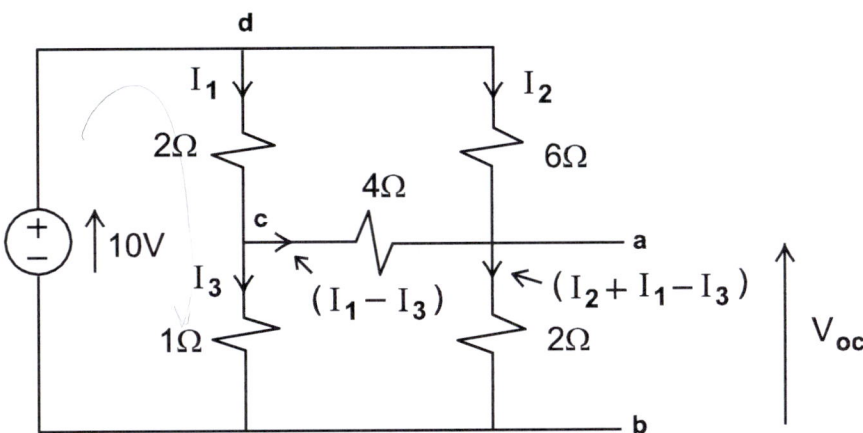

Fig. 1.34 Circuit for finding the Thévenin equivalent voltage (Example 1.16).

$$10 = 2I_1 + I_3 \qquad (1.68)$$

$$10 = 2I_1 + 4(I_1 - I_3) + 2(I_2 + I_1 - I_3)$$
$$10 = 8I_1 + 2I_2 - 6I_3 \qquad (1.69)$$

$$10 = 6I_2 + 2(I_2 + I_1 - I_3)$$
$$5 = I_1 + 4I_2 - I_3 \qquad (1.70)$$

Solving Equs. 1.68 to 1.70 simultaneously gives

$$I_1 = 3.38 \text{ A}$$
$$I_2 = 1.22 \text{ A}$$
$$I_3 = 3.24 \text{ A} \qquad (1.71)$$

Then

$$V_{oc} = V_{Th} = 2(1.22 + 3.38 - 3.24)$$
$$V_{Th} = 2.70 \text{ V} \qquad\qquad (1.72)$$

b) I_N

I_N is the short-circuit current that flows between a and b when these terminals are short circuited (see the circuit in Fig. 1.35).

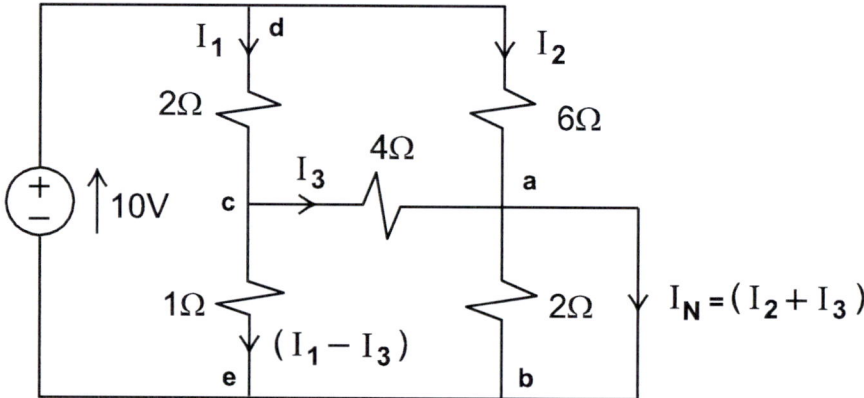

Fig. 1.35 Circuit for finding the Norton equivalent current (Example 1.16).

Because of the short circuit, the voltage across the 2Ω resistor (connected between a and b) is zero. Then 10 volts appear across the 6Ω resistor and Ohm's law gives

$$I_2 = (10/6)$$
$$I_2 = 1.67 \text{ A} \qquad\qquad (1.73)$$

10 volts also appear across each of the paths dca and dce. KVL then gives

$$10 = 2I_1 + 4I_3 \tag{1.74}$$
$$10 = 3I_1 - I_3 \tag{1.75}$$

Solving Equs. 1.74 and 1.75 simultaneously gives

$$I_1 = 3.57\,\text{A} \tag{1.76}$$
$$I_3 = 0.71\,\text{A} \tag{1.77}$$

Then

$$I_N = I_2 + I_3 \tag{1.78}$$
$$I_N = 2.38\ \text{A} \tag{1.79}$$

c) R_{Th}

$$R_{Th} = \frac{V_{oc}}{I_N} \tag{1.80}$$
$$R_{Th} = 1.13\ \Omega \tag{1.81}$$

This can be verified by finding the equivalent resistance looking left into terminals a and b, in the circuit of Fig. 1.36. The circuit of Fig. 1.36 shows the circuit of Fig. 1.33 with the voltage source in the latter circuit deactivated. It can be seen that the 2Ω and the 1Ω resistors are in parallel with one another, and

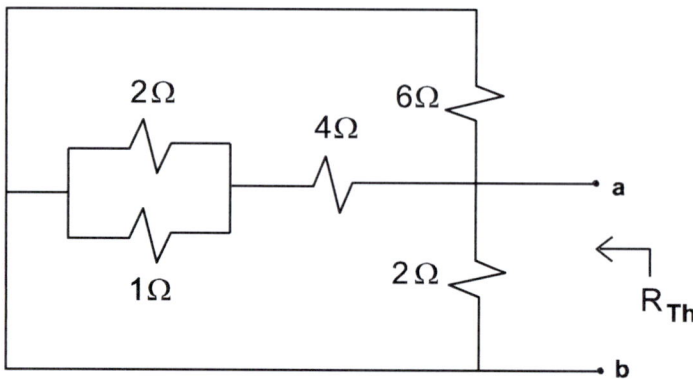

Fig. 1.36. Circuit for finding the Thévenin equivalent resistance (Example 1.16).

that their equivalent resistance, in series with the 4Ω resistor, is in parallel with the 6Ω and the 2Ω resistors. Therefore,

$$R_{Th} = \cfrac{1}{(1/6)+(1/2)+\left\{\cfrac{1}{4+[1/(1/2+1/1)]}\right\}} \qquad (1.82)$$

Equation 1.82 gives the same result of 1.13Ω.

Example 1.17: Thévenin and Norton equivalents

Find V_{Th}, I_N and R_{Th} for the circuit to the left of terminals a and b in Fig. 1.37.

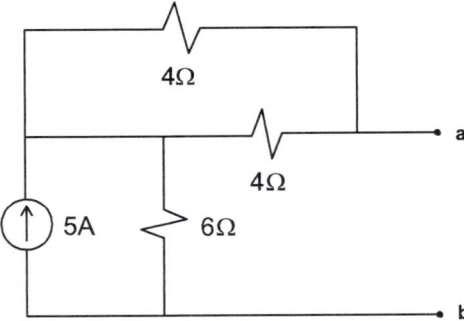

Fig. 1.37 Find the Thévenin and Norton equivalent circuit parameters looking left into terminals a and b (Example 1.17).

a) V_{Th}

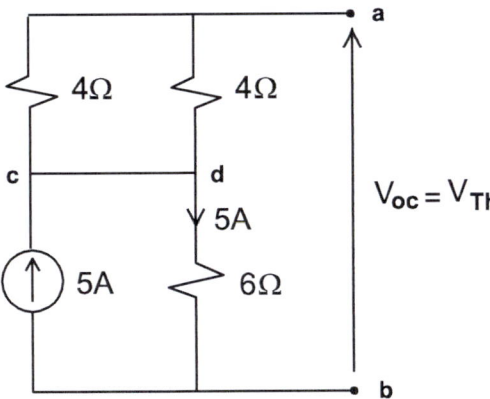

Fig. 1.38 Circuit for finding V_{Th} (Example 1.17).

V_{Th} is equal to the open circuit voltage V_{oc} in Fig 1.38. The current of 5A entering node c encounters an 8Ω resistance (the

43

series combination of the two 4Ω resistors) in parallel with a short circuit between points *c* and *d*. Since the short circuit has zero resistance, this current completely bypasses the two 4Ω resistances and flows, first through the short circuit and then through the 6Ω resistor. Then

$$V_{oc} = V_{ab} = V_{Th} = 5 \times 6 = 30 \text{ V} \tag{1.83}$$

b) I_N and R_{Th}

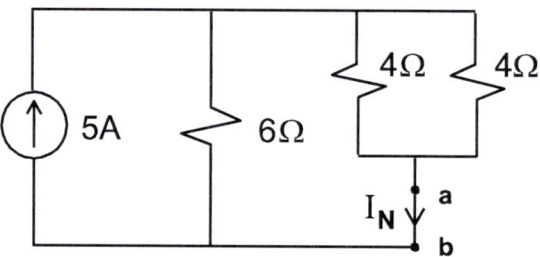

Fig. 1.39 Circuit for finding I_N (Example 1.17).

The Norton current is the current flowing between *a* and *b*, when these terminals are short-circuited together, as shown in Fig. 1.39. Redrawing Fig. 1.37, as in Fig. 1.39, shows that I_N is the current that flows through the parallel equivalent of the two 4Ω resistors. Since the 5A supplied by the current source is divided between this 2Ω equivalent resistor and the 6Ω resistor, the current division equation gives

$$I_N = \left(\frac{1/2}{1/2 + 1/6} \right) 5 = 3.75 \tag{1.84}$$

$$R_{Th} = V_{Th} / I_N = 30/3.75 = 8\Omega \tag{1.85}$$

1.11 Node voltage method

The node voltage method can be used to analyze circuits with multiple nodes. It is a powerful technique for solving for unknown voltages in circuits where the methods discussed earlier may not be fruitful.

Consider a circuit with multiple nodes. Each node has a voltage with respect to another node. In the node voltage method each node in the circuit is first identified. Secondly, one of these nodes is chosen as the reference node and each remaining node is assigned a voltage with respect to the reference node. These voltages are then referred to as the node voltages. Thirdly, where necessary and possible, using Ohm's law or KVL, branch currents are expressed in terms of node voltages. Fourthly, KCL is applied to sum currents at nodes. This results in a set of equations involving the node voltages, where solving the equations simultaneously gives the node voltages. Branch currents and power values can then be calculated.

Example 1.18: Node voltage method

Find V_{ao} and V_{bo} in the circuit of Fig. 1.40.

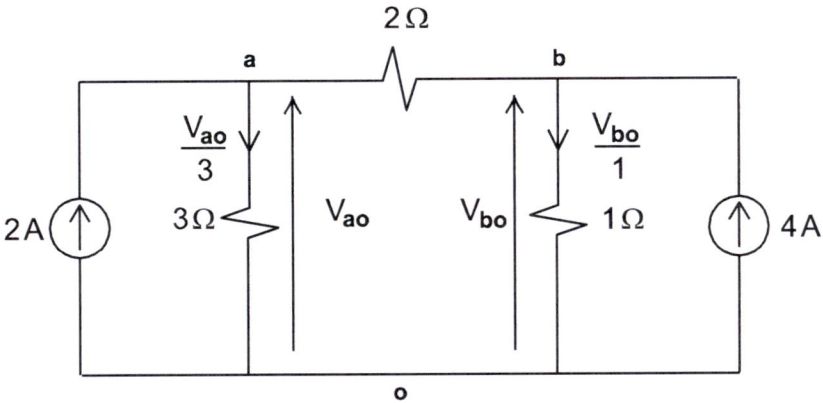

Fig. 1.40 A three-node circuit (Example 1.18).

In this circuit o is chosen as the reference node and the voltages of the remaining nodes, a and b, are assumed to be with respect to o. Then V_{ao} and V_{bo} are the node voltages. Assuming $V_{ao} > V_{bo}$ then current is assumed to flow from node a to b, and KCL applied at nodes a and b gives

$$2 = \frac{V_{ao}}{3} + \frac{(V_{ao} - V_{bo})}{2} \tag{1.86}$$

$$4 + \frac{(V_{ao} - V_{bo})}{2} = \frac{V_{bo}}{1} \tag{1.87}$$

Equations 1.86 and 1.87 give

$$5V_{ao} - 3V_{bo} = 12 \tag{1.88}$$

$$V_{ao} - 3V_{bo} = -8 \tag{1.89}$$

Solving simultaneously gives the node voltages as $V_{ao}=5V$ and $V_{bo}=4.33V$.

If the 2Ω resistor were replaced with a short circuit, in the circuit of Fig. 1.34, then the circuit would have two nodes and could be analyzed using the single-node-pair method discussed in section 1.6.

Example 1.19: Node voltage method

Find I_1 and I_2 in the circuit of Fig. 1.41.

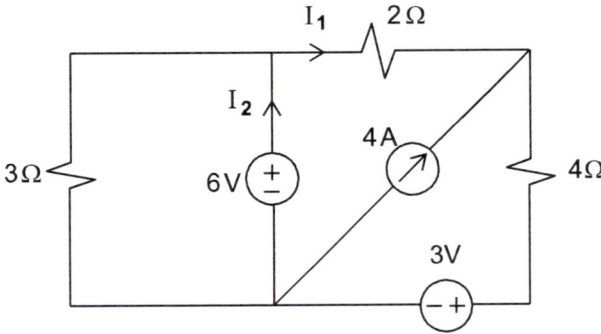

Fig. 1.41 A four-node circuit to be analyzed using the node voltage method (Example 1.19).

Figure 1.42 shows a redrawn version of the circuit of Fig. 1.41.

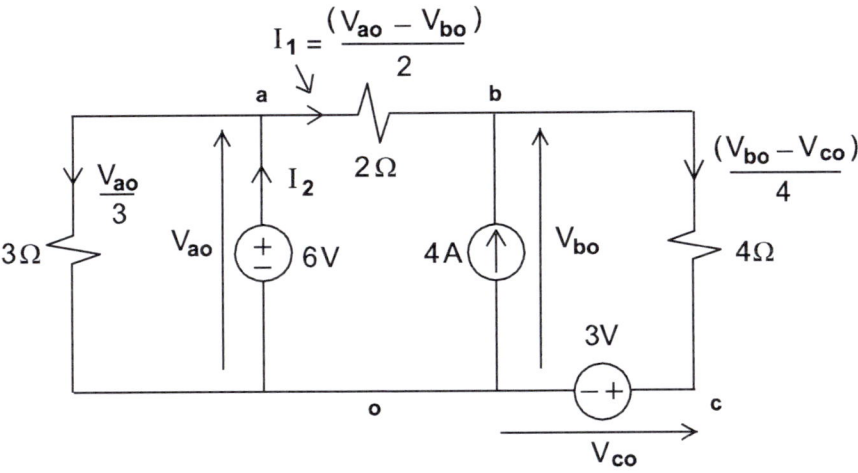

Fig. 1.42 The circuit of Fig. 1.41 set up for node voltage analysis (Example 1.19).

There are four nodes o, a, b and c and the node voltages are V_{ao}, V_{bo} and V_{co}. Node o is chosen as the reference node. Inspection of the circuit shows that the voltage sources ensure that the potential of node a is 6V above that of the reference node and node c is 3V above node o. Therefore

$$V_{ao} = 6 \text{ V} \tag{1.90}$$

$$V_{co} = 3 \text{ V} \tag{1.91}$$

V_{ao} is the voltage difference appearing across the terminals of the 3Ω resistor. The voltage difference across the 2Ω and 4Ω resistors is $(V_{ao} - V_{bo})$ and $(V_{bo} - V_{co})$, respectively. Then Ohm's law gives the expressions for the currents flowing through these resistors and KCL, applied at nodes a and b, gives

$$\frac{V_{ao}}{3} + \frac{(V_{ao} - V_{bo})}{2} = I_2 \tag{1.92}$$

$$\frac{(V_{ao} - V_{bo})}{2} + 4 = \frac{(V_{bo} - V_{co})}{4} \tag{1.93}$$

Substituting for V_{ao} and V_{co} from Equs. 1.90 and 1.91 into 1.92 and 1.93, and solving the last two equations simultaneously gives

$$V_{bo} = 10.33 \text{ V} \tag{1.94}$$

$$I_2 = -0.17 \text{ A} \tag{1.95}$$

$$I_1 = \frac{(6 - 10.33)}{2} = -2.17 \text{ A} \tag{1.96}$$

Example 1.20: Node voltage method and supernode

Find V_{ao} and V_{bo} in the circuit of Fig. 1.43.

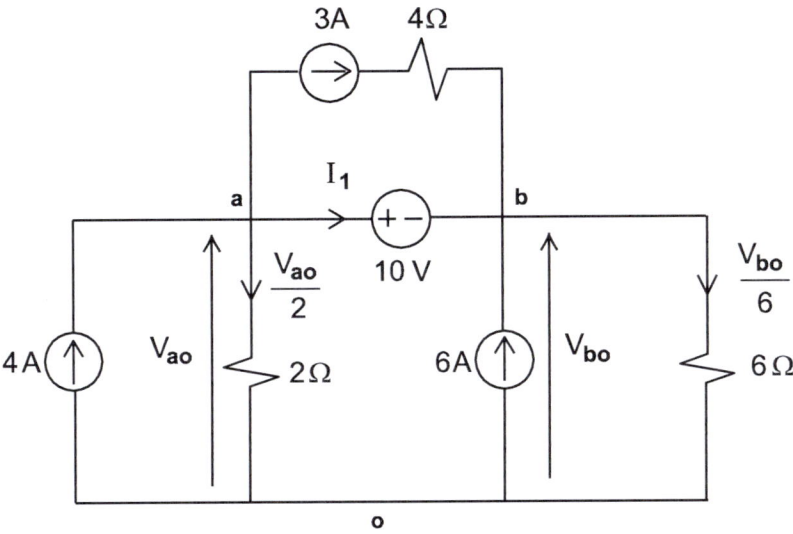

Fig. 1.43 Connecting the voltage source between the two non-reference nodes a and b results in a supernode (Example 1.20).

The reference node is o, in Fig. 1.43, and the voltages of the non-reference nodes a and b, with respect to o, need to be found. As before, the voltage arrows representing these voltages V_{ao} and V_{bo} are drawn between the nodes and the reference point. Expressions obtained using Ohm's law, for branch currents flowing through the 2Ω and 6Ω resistors, are shown next to the current arrows.

Although KCL can now be applied to nodes a and b, there is an alternative to this. It is realized that nodes a and b constitute a single node, referred to as a supernode, as far as the application of KCL is concerned. A supernode is obtained when a voltage source is connected between two non-reference nodes. For example, the 10V source that is connected between nodes a and b results in a supernode. In applying KCL to a supernode the two non-reference nodes are assumed to be one node. A current such as I_1, or the 3A source current, flowing out of one non-reference node and into the other non-reference node, within the supernode, is ignored. KCL is then applied to currents external to the supernode so that the sum of currents (4A+6A) flowing into the supernode, in Fig. 1.43, equals the sum flowing out. This gives Equ. 1.97.

$$4 + 6 = \frac{V_{ao}}{2} + \frac{V_{bo}}{6} \qquad (1.97)$$

(This is easily shown to be true by applying KCL at nodes a and b which gives

$$4 - \frac{V_{ao}}{2} = 3 + I_1 \qquad (1.98)$$

$$\frac{V_{bo}}{6} - 6 = 3 + I_1 \qquad (1.99)$$

Equating the left hand sides of Equs. 1.98 and 1.99 gives Equ. 1.97).

KVL applied to loop abo gives

$$V_{ao} - 10 - V_{bo} = 0 \qquad (1.100)$$

Solving Equs. 1.97 and 1.100 simultaneously gives

$$V_{ao} = 17.5V \qquad (1.101)$$
$$V_{bo} = 7.5V \qquad (1.102)$$

This shows that both nodes a and b are at potentials above the reference potential. Also, the currents flowing through the 2Ω and 6Ω resistors flow from nodes a and b to the reference node, as assumed in Fig. 1.43.

Example 1.21: Node voltage method

With node *o* chosen as the reference, in the circuit of Fig. 1.44, find the voltages of nodes *a*, *b* and *c* with respect to the reference.

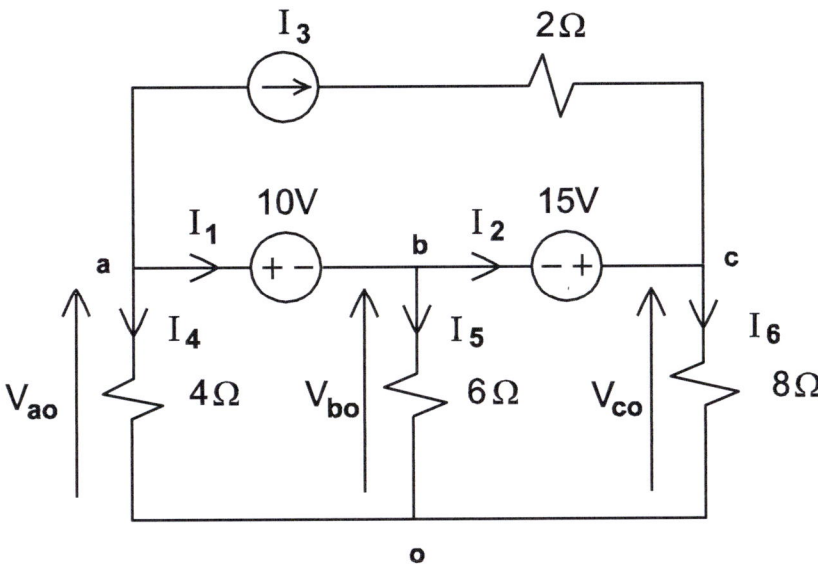

Fig. 1.44 Nodes *a*, *b* and *c* are part of a supernode (Example 1.21).

It has been assumed that the potential at each of nodes *a*, *b* and *c* is above reference and, consequently, current I_4, I_5, and I_6 are shown flowing away from these three nodes towards the reference node. KCL at nodes *b* and *c* gives

$$I_5 = I_1 - I_2 \qquad\qquad (1.103)$$
$$I_6 = I_2 + I_3 \qquad\qquad (1.104)$$

Adding Equs. 1.103 and 1.104 gives

51

$$I_5 + I_6 = I_1 + I_3 \qquad (1.105)$$

KCL at node a gives

$$I_4 = -(I_1 + I_3) \qquad (1.106)$$

Comparing Equs. 1.105 and 1.106 gives

$$I_4 = -(I_5 + I_6) \qquad (1.107)$$
$$I_4 + I_5 + I_6 = 0 \qquad (1.108)$$

Equ. 1.108 is an expression of KCL at the supernode to which nodes a, b and c belong. This shows that more than two nodes can form a supernode. Using Ohm's law to express the currents in Equ. 1.108 in terms of node voltages gives

$$\frac{V_{ao}}{4} + \frac{V_{bo}}{6} + \frac{V_{co}}{8} = 0 \qquad (1.109)$$
$$6V_{ao} + 4V_{bo} + 3V_{co} = 0 \qquad (1.110)$$

Using KVL in loops abo and bco gives

$$V_{ao} - V_{bo} = 10 \qquad (1.111)$$
$$V_{bo} - V_{co} = -15 \qquad (1.112)$$

Solving the last three equations simultaneously gives

$$V_{ao} = 1.9\,\text{V} \qquad (1.113)$$
$$V_{bo} = -8.08\,\text{V} \qquad (1.114)$$
$$V_{co} = 6.92\,\text{V} \qquad (1.115)$$

1.12 Mesh current method

The mesh current method, like the node voltage method, is used to solve for unknown voltages and current in a circuit. In the mesh current method a current is associated with each mesh of the circuit. A loop is a path that starts and ends at the same node. A mesh is a loop that does not enclose other loops. The path *abcdefa*

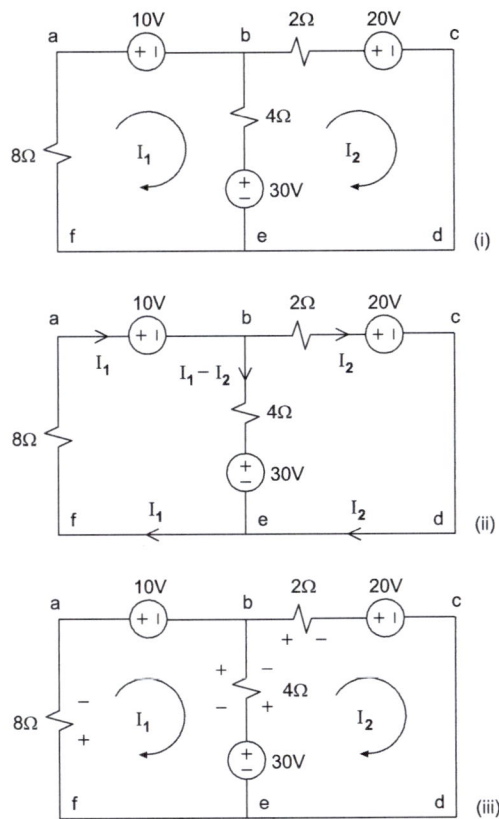

Fig 1.45 Circuit for mesh current analysis.

in Fig. 1.45(i) is a loop but not a mesh, whereas the path *abefa* is a loop and a mesh. Mesh current are assumed here to flow in a clockwise direction and are indicated as I_1 and I_2 in the circuit of

Fig. 1.45(i). The three circuits in Fig. 1.45 are the same. The circuit of Fig. 1.45(ii) shows specific branch currents in terms of the mesh currents. The current through the path ab, for instance, is the mesh current I_1. The current through the 4Ω resistor is $I_1 - I_2$, because this resistor is shared by both meshes. Therefore, resistors that are common to two meshes carry two components of current, each component being a mesh current, and consequently, each component of current results in a voltage drop the polarity of which depends on the direction of the individual mesh current. In the mesh current method KVL is applied to each mesh. This results in a set of simultaneous equations that are solved to determine the mesh currents. In applying KVL it is useful to represent the polarity of voltages across resistors as shown in Fig. 1.45(iii). Here the shared resistor is shown with two voltages across it so that the total voltage drop is the algebraic sum of the two voltages.

KVL applied to the two meshes in the circuit of Fig. 1.45 gives

$$8I_1 + 30 - 4I_2 + 4I_1 + 10 = 0 \tag{1.116}$$
$$4I_2 - 4I_1 + 2I_2 + 20 - 30 = 0 \tag{1.117}$$

These give the simultaneous equations

$$(8+4)I_1 - 4I_2 = -40 \tag{1.118}$$
$$-4I_1 + (4+2)I_2 = 10 \tag{1.119}$$

Solving these simultaneously, by calculator, gives $I_1 = -3.57\,\text{A}$, and $I_2 = -0.71\,\text{A}$.

In Equ. 1.118, for mesh 1, the coefficient of I_1 is the sum of the resistances of the resistors in mesh 1, whereas the coefficient of I_2 is the resistance of the resistor common to meshes 1 and 2. The right hand side of the equation is the algebraic sum of voltage sources in mesh 1. The sign of a voltage is positive if this voltage source, acting as a generator, would give a current that flows in the same direction as the mesh current.

Example 1.22: Mesh current method

Determine the mesh currents in the circuit of Fig. 1.46.

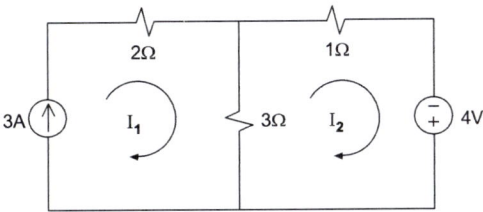

Fig. 1.46 Circuit for Example 1.22.

From mesh 1, $I_1 = 3\,\mathrm{A}$. For mesh 2
$$-3I_1 + (3+1)I_2 = 4$$
$$I_2 = 3.25\,\mathrm{A}$$

Example 1.23: Mesh current method

Determine the mesh currents in the circuit of Fig. 1.47.

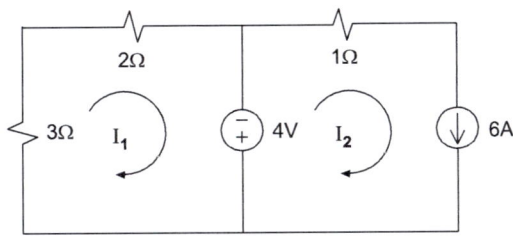

Fig. 1.47 Circuit for Example 1.23.

From mesh 2, $I_2 = 6\,\mathrm{A}$. For mesh 1
$$(3+2)I_1 - 0 \times I_2 = 4$$
$$I_1 = 0.8\,\mathrm{A}.$$

Example 1.24: Mesh current method

Determine the mesh currents in the circuit of Fig. 1.48.

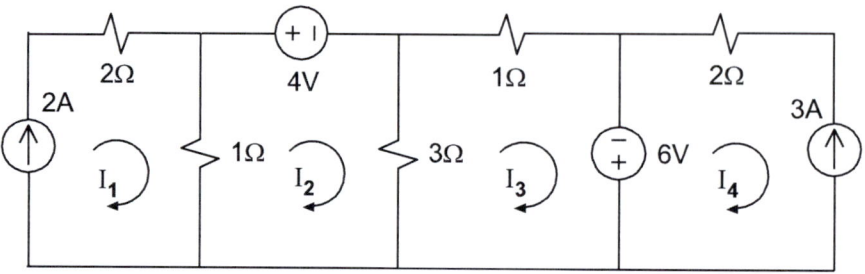

Fig. 1.48 Circuit for Example 1.24.

From mesh 1, $I_1 = 2\,\text{A}$, and from mesh 4, $I_4 = -3\,\text{A}$. KVL for mesh 2 gives

$$-1 \times I_1 + (1+3)I_2 - 3I_3 - 0 \times I_4 = -4$$
$$4I_2 - 3I_3 = -2$$

KVL for mesh 3 gives

$$0 \times I_1 - 3I_2 + (3+1)I_3 - 0 \times I_4 = 6$$
$$-3I_2 + 4I_3 = 6$$

Solving the two simultaneous equations gives

$$I_2 = 1.43\,\text{A, and } I_3 = 2.57\,\text{A}.$$

1.13 Maximum power transfer to load

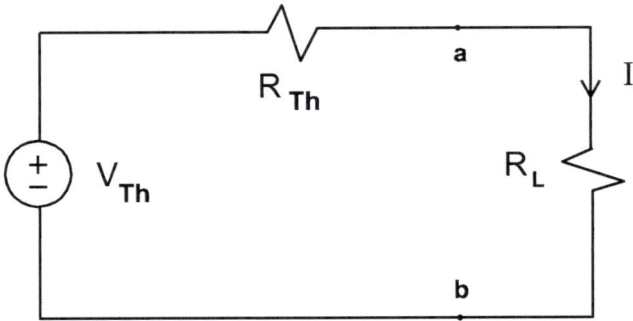

Fig. 1.49 An electrical system, represented by its Thevenin equivalent circuit, supplies power to load resistor R_L.

The circuit to the left of terminals a and b, in Fig. 1.49, represents a source supplying electrical power to the load resistor R_L. The amount of power transferred from the source to the load is important. If the source represents a power utility system involving the generation, transmission and distribution of electrical power, then the amount of power delivered to a load is important, as an inefficient transfer of power means the loss of power in transmission and distribution. If the circuit represents an instrumentation or communication system that is transferring information through electrical signals, then it is important to maximize the power received by the load.

The system to the left of a and b can be represented by its Thevenin equivalent circuit, as shown in Fig. 1.49. For a given system the parameters of the Thevenin equivalent circuit are fixed. The value of R_L can be changed, however, to find a load resistance for which the power transferred to it is a maximum. To determine R_L, the expression of the power P that is delivered to the resistor is considered, where

$$P = I^2 R_L \qquad\qquad (1.120)$$

57

Substituting the expression for the current into Equ. 1.120 gives

$$P = \frac{V_{Th}^2}{(R_{Th} + R_L)^2} R_L \tag{1.121}$$

To find the value of the load resistor that maximizes the power delivered to the load, the derivative of P with respect to R_L is found, and then set to zero

$$\frac{dP}{dR_L} = \frac{V_{Th}^2 (R_{Th} + R_L)^2 - 2V_{Th}^2 R_L (R_{Th} + R_L)}{(R_{Th} + R_L)^4} = 0 \tag{1.122}$$

$$(R_{Th} + R_L)^2 = 2R_L (R_{Th} + R_L) \tag{1.123}$$

$$R_L = R_{Th}. \tag{1.124}$$

Therefore, maximum power transfer takes place when the load resistor is equal to the Thevenin resistance. Substituting for R_{Th} from Equ. 1.124, into Equ.1.121, gives the maximum power transferred to the load as

$$P_{max} = \frac{V_{Th}^2}{4R_L} . \tag{1.125}$$

Chapter 2
Capacitor and inductor

2.1 Introduction

The principles and methods considered in Chapter 1 were illustrated in relation to resistive circuits. Two additional components, the capacitor and the inductor, are now introduced. Whereas the resistor is a dissipative element, where electrical energy is always irreversibly converted to heat, the capacitor and inductor can store energy, which can then be released to the circuit. For this reason they are referred to as energy storage elements. Their use results in circuits of increased signal processing capabilities so that a filter in a radio, for instance, can select a particular station frequency, electrical power can be efficiently delivered to the consumer, and an electronic flash unit in a camera can convert stored electrical energy into a burst of light.

The current-voltage relationships for the capacitor and inductor and their relevance to methods of analysis of capacitor and inductor circuits under dc (constant current and voltage) conditions will be considered in this chapter. These relationships will also be necessary in later chapters, to describe the capacitor and inductor under ac (alternating current) conditions. The energy stored in an inductor or capacitor cannot change instantaneously; this principle will be used to determine the initial value of current for an inductor and of voltage for a capacitor, immediately after a switch is opened or closed in a circuit. This method will be applied in Chapter 6 in the analysis of transient currents and voltages. Transients describe the variation of currents and voltages after a switch is opened or closed, as a circuit goes from one steady-state (constant current and voltage) condition to another.

2.2 Capacitor current, voltage and energy relationships
2.2.1 Charge-voltage relationship

Two overlapping conducting plates separated by an insulator result in a capacitor. Consider the capacitor shown in Fig. 2.1 where two metal plates a and b, each of area A, constitute the two electrodes of the capacitor. The distance between the electrodes is d. Assuming that the capacitor has no initial charge on its plates, then if at time $t=0$ a voltage source is connected in series with the capacitor, as in the circuit of Fig. 2.1, positive charge starts flowing to electrode a and an equal amount of negative charge flows at the same rate to b. The transfer of charge is accompanied by an increase in the voltage across the capacitor so that over a period of time the voltage across the capacitor increases from an initial value of zero to the value of the supply voltage V_s. The transfer of charge ceases when the voltage across the capacitor equals the supply voltage. The capacitor is then fully charged.

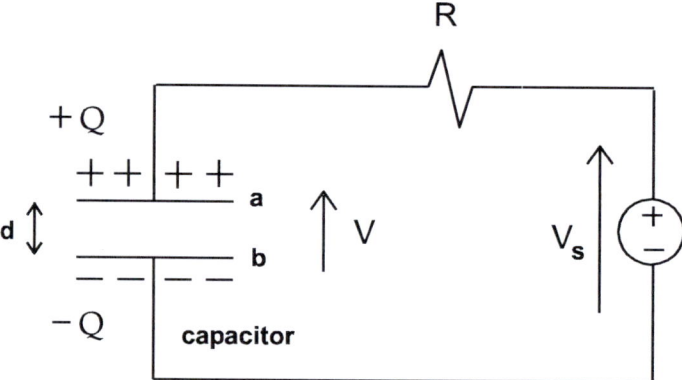

Fig. 2.1 A parallel plate capacitor, with no initial charge on its metal plates a and b for time t <0, is connected to a voltage source V_s at time $t=0$. The circuit shows conditions at time t >0.

60

If Q coulombs is the magnitude of the charge residing on a plate then

$$Q = CV \qquad (2.1)$$

where V is the voltage across the capacitor and C is the capacitance in farads (abbreviated F). A capacitor has a capacitance of one farad when a voltage of one volt across its electrodes results in a charge of magnitude one coulomb residing on each plate. A farad is a large value of capacitance; typical capacitor values are in the pF (picofarad=10^{-12} F) to µF (microfarad=10^{-6} F) range.

The capacitance of a capacitor can be increased by having insulating materials (dielectrics) between the plates that have permittivities higher than air. For the parallel plate capacitor of Fig. 2.1 the capacitance is given by

$$C = \frac{\varepsilon_o \varepsilon_r A}{d} \qquad (2.2)$$

where ε_o= permittivity of free space (8.85×10^{-12} farads/meter) and ε_r= relative permittivity of the dielectric. The relative permittivity of air is approximately one while that of nylon 3.5, of corundum (Al_2O_3 alumina) 10, and of rutile (TiO_2 titania) 110. For a given voltage applied across the capacitor, the higher the relative permittivity, higher the charge stored on the plates of the capacitor.

2.2.2 Current-voltage relationship for capacitor

The rate of flow of charge to the plates of a capacitor results in a current so that the current at time t is

$$i(t) = \frac{dq(t)}{dt} \qquad (2.3)$$

where $q(t)$ is the charge on a plate at time t.

Differentiating Equ. (2.1) gives

$$i(t) = C\frac{dv(t)}{dt} \qquad (2.4)$$

This is the current-voltage relationship for the capacitor where $v(t)$ is the voltage across the capacitor at t. Current flows in the direction of positive charge. The capacitor plate that current flows into is the more positive electrode and the voltage arrow points towards it, as shown in Fig. 2.2.

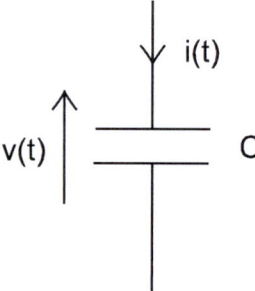

Fig. 2.2 The electrical symbol for the capacitor of capacitance C farads.

As the capacitor charges from time $t=0$ to a general time t, the voltage across it increases to $v(t)$, where this voltage is obtained by integrating Equ. (2.4) to get

$$v(t) = \frac{1}{C}\int_0^t i(t)dt \qquad (2.5)$$

If an initial voltage appears across the inductor at $t=0^-$, the value given by Equ. (2.5) is added to the initial voltage to find the voltage at t.

2.2.3 Energy stored in capacitor

The power delivered to the capacitor at time t is given by

$$p(t) = v(t)i(t) \qquad (2.6)$$

Substituting for the current from Equ. (2.4) into (2.6) gives

$$p(t) = Cv(t)\frac{dv(t)}{dt} \qquad (2.7)$$

Since $p(t)$ is the energy delivered to the capacitor per second at time t, then the total energy stored in the capacitor as charge is delivered to its plates from time zero to t is found by integrating $p(t)$ in Equ. (2.7) between these time limits so that the energy is

$$U = \int_0^t p(t)dt \qquad (2.8)$$

$$U = C\int_0^t v(t)\frac{dv(t)}{dt}dt \qquad (2.9)$$

$$U = \frac{1}{2}Cv^2(t) - \frac{1}{2}Cv^2(0) \qquad (2.10)$$

where

$$U = \frac{1}{2}Cv^2(0) \qquad (2.11)$$

is the initial energy stored in the capacitor, where it is assumed that at $t=0$ the voltage across the capacitor is $v(0)$.

In general, charging a capacitor, so that the voltage across its plates becomes V, results in the storage of

$$U = \frac{1}{2}CV^2 \qquad\qquad (2.12)$$

joules of energy in the electric field between the plates of the capacitor. The electric field is given by

$$E = \frac{V}{d} \text{ volts per meter.} \qquad\qquad (2.13)$$

Energy cannot change instantaneously and since Equ. (2.12) shows that the energy stored in a capacitor is equal to the square of the voltage, then the voltage across the capacitor cannot change instantaneously.

Example 2.1: Initial and final conditions in RC circuit

In the circuit of Fig. 2.3(i) the switch S is open for time t<0 and during this time the capacitor is uncharged. At $t = 0$ the switch is closed. Find V_1, V_2 and I immediately before the switch is closed at $t = 0^-$, immediately after the switch is closed at $t = 0^+$, and when the switch has been closed for a long time $t = \infty$, to complete the table in Fig. 2.3 (ii).

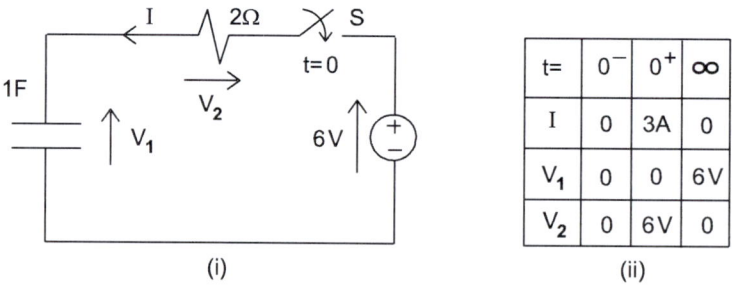

t=	0^-	0^+	∞
I	0	3A	0
V_1	0	0	6V
V_2	0	6V	0

(i) (ii)

Fig. 2.3 Find initial and final values of V_1, V_2 and I (Example 2.1).

64

At $t<0$ the switch is open and $I=0$. No current flows through the resistor and so $V_2=0$. Since the capacitor carries no charge on its plates during this time $V_1=0$. The voltage across the capacitor, both immediately before the switch is closed and immediately after, is the same. This is due to the fact that the voltage across the capacitor cannot change instantaneously. Therefore, $V_1=0$ at $t=0^-$ and at $t=0^+$.

At $t=0^+$ KVL gives

$$V_1 + V_2 = 6 \tag{2.14}$$

and since $V_1=0$ the 6V supply voltage appears entirely across the 2Ω resistor, $V_2=6$V and from Ohm's law $I=(6/2)=3$A. Although the voltage across the capacitor is 0 at $t=0^+$ the current through the capacitor is at its maximum and the rate of increase of the voltage across the capacitor, from Equ.2.4, is at its maximum.

As time increases the charge flowing to the capacitor gradually increases the voltage across the capacitor (since $Q=CV$). As V_1 increases, the voltage across the resistor, $(6-V_1)$, decreases and from Ohm's law so does the current though the resistor and consequently through the capacitor. When V_1 reaches the supply voltage of 6V the voltage drop across the resistor is zero and the current stops. Therefore at $t=\infty$ $V_1=6$V, $I=0$ and $V_2=0$.

In summary, immediately after the switch is closed, the voltage across the capacitor stays at the same value it had immediately before the switch was closed; the current through the capacitor, however, jumps from zero to its maximum value. As time increases, the voltage across the capacitor gradually increases until it equals the voltage of the voltage source and the current through the capacitor gradually decreases to zero. In Chapter 6, the time constant which characterizes the increase or decrease of these transient variations of voltage and current, in the time range of $t=0^+$ to $t=\infty$, will be discussed.

2.3 Combining capacitors in series and parallel
2.3.1 Parallel combination

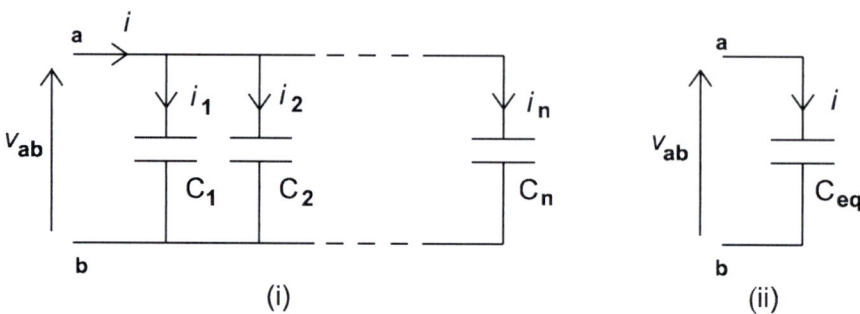

Fig. 2.4 Capacitors in parallel.

The parallel combination of capacitors between terminals a and b in Fig. 2.4(i) can be replaced with one equivalent capacitor C_{eq} as in (ii). Using KCL

$$i = i_1 + i_2 + ... + i_n \qquad (2.15)$$

$$i = (C_1 + C_2 + ... + C_n)\frac{dv}{dt} \qquad (2.16)$$

$$i = C_{eq}\frac{dv}{dt} \qquad (2.17)$$

$$C_{eq} = C_1 + C_2 + ... + C_n \qquad (2.18)$$

Equ. (2.18) shows that capacitors connected in parallel combine as resistors connected in series.

2.3.2 Capacitors in series

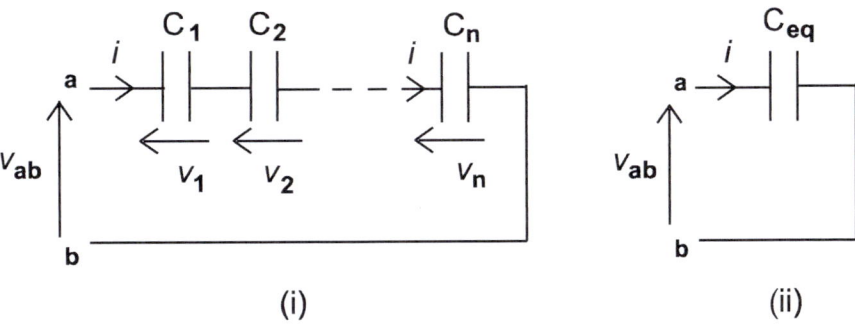

Fig. 2.5 Capacitors connected in series.

Assuming the capacitors in Fig. 2.5(i) are uncharged at $t=0$ and so the voltage across individual capacitor is zero at that time, then as current i starts to flow at $t=0$, the voltage across each capacitor starts to increase so that at time t the voltage between terminals a and b is, from KVL and Equ. (2.5), given by

$$v(t) = \frac{1}{C_1}\int_0^t i(t)dt + \frac{1}{C_2}\int_0^t i(t)dt + ... + \frac{1}{C_n}\int_0^t i(t)dt \qquad (2.19)$$

$$v(t) = (\frac{1}{C_1} + \frac{1}{C_2} + ... + \frac{1}{C_n})\int_0^t i(t)dt \qquad (2.20)$$

$$v(t) = \frac{1}{C_{eq}}\int_0^t i(t)dt \qquad (2.21)$$

$$C_{eq} = \frac{1}{(1/C_1 + 1/C_2 + ... + 1/C_n)} \qquad (2.22)$$

Equ. (2.22) shows that capacitors in series combine as resistors in parellel.

67

Example 2.2: Charging capacitor

A constant current of 4A flows into a previously uncharged 100μF capacitor for 2ms (millisecond=10^{-3} second). Find the voltage difference between the plates of the capacitor at the end of the charging period.

If the charge on a capacitor plate at time t is q then the current flowing into the capacitor at t is

$$i = \frac{dq}{dt} \tag{2.23}$$

and the charge that resides on a plate after charging for t seconds is

$$q = \int_0^t i\,dt \tag{2.24}$$

Since the charging current is a constant I, then from Equ. 2.24, the charge on a plate at t is

$$Q = It = 4 \times 2 \times 10^{-3} = 8 \times 10^{-3} \text{ C} \tag{2.25}$$

Then, from Equ. (2.1), the voltage difference between the plates at the end of charging is

$$V = \frac{Q}{C} = \frac{8 \times 10^{-3}}{100 \times 10^{-6}} = 80 \text{ V} \tag{2.26}$$

Example 2.3: Discharging capacitor

A 47μF capacitor has been charged so that 100V appears across its plates. The capacitor now needs to be connected across a resistor so that it provides an average current of 5mA to flow through the resistor. How long can the capacitor maintain this current?

The charge stored in the capacitor is

$$Q = CV = 47 \times 10^{-6} \times 100 = 4.7 \times 10^{-3} \text{ C} \qquad (2.27)$$

The stored charge can be used to provide an average discharge current of I, for time t, where from $Q = It$

$$t = \frac{4.7 \times 10^{-3}}{5 \times 10^{-3}} = 0.94 \text{ seconds} \qquad (2.28)$$

Example 2.4 Charge stored in capacitors connected in parallel

Find the equivalent capacitance between terminals a and b, the charge stored in this capacitor and the charge stored in the 1μF and 2μF capacitors.

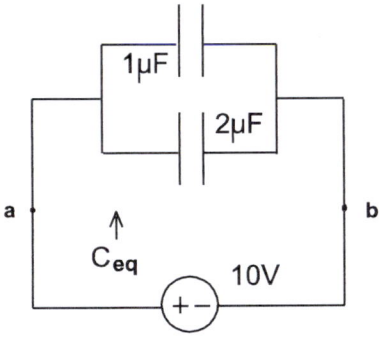

Fig. 2.6 Capacitors in parallel (Example 2.4).

The two capacitors combine in parallel to give an equivalent

$$C_{eq} = (1+2)10^{-6} = 3 \times 10^{-6} \text{ F} \qquad (2.29)$$

The charge stored on C_{eq} is

$$Q = C_{eq}V = 3 \times 10^{-6} \times 10 = 30 \ \mu\text{C} \qquad (2.30)$$

The charge stored on the $1\mu\text{F}$ capacitor is

$$Q_1 = 1 \times 10^{-6} \times 10 = 10 \ \mu\text{C} \qquad (2.31)$$

The charge stored on the $2\mu\text{F}$ capacitor is

$$Q_2 = 2 \times 10^{-6} \times 10 = 20 \ \mu\text{C} \qquad (2.32)$$

The results show that for capacitors connected in parallel, the charge stored on the equivalent capacitor is equal to the sum of the charges stored on the individual capacitors.

Example 2.5: Charge stored in capacitors connected in series

Find the equivalent capacitance of the two capacitors, the charge on the equivalent and the individual capacitors and the voltage across each individual capacitor.

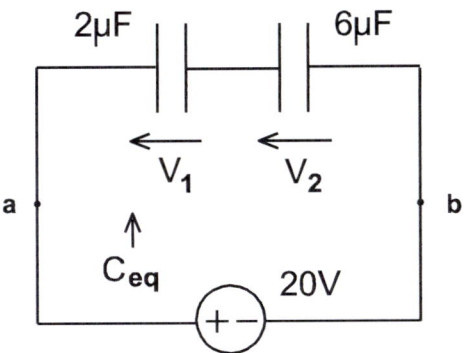

Fig. 2.7 Capacitors in series (Example 2.5).

Combining the two capacitors in series gives the equivalent capacitor as

$$C_{eq} = \frac{1}{(1/2\times10^{-6})+(1/6\times10^{-6})} = 1.5 \ \mu F \tag{2.33}$$

The charge on C_{eq} is

$$Q = C_{eq} \times 20 = 1.5\times10^{-6} \times 20 = 30 \ \mu C \tag{2.34}$$

The charge on each capacitor in series is the same and also equal to the charge on the equivalent capacitor. The circuit of Fig. 2.7 depicts a steady state situation where the two capacitors are fully charged and no current flows. When these capacitors, initially uncharged, were first connected to the supply voltage, a charging current started circulating in the circuit and, because the capacitors were in series, the current flowing through each capacitor was the same. Since the current is the flow of charge to a capacitor, equal current means an equal amount of charge was delivered to each capacitor. Therefore, when the charging was completed and the current stopped flowing, an equal amount of charge resided on each capacitor.

Therefore, the voltage across the 2μF capacitor is

$$V_1 = \frac{Q}{C_1} = \frac{30 \times 10^{-6}}{2 \times 10^{-6}} = 15 \text{ V} \tag{2.35}$$

and V_2 is given by

$$V_2 = \frac{Q}{C_2} = \frac{30 \times 10^{-6}}{6 \times 10^{-6}} = 5 \text{ V} \tag{2.36}$$

where $(V_1 + V_2)$ equals the 20V of the voltage source, as expected from KVL.

Example 2.6:   Charge stored on capacitors in series and parallel under steady-state conditions

Find the voltage across and the charge on, each capacitor in the circuit of Fig. 2.8 (i).

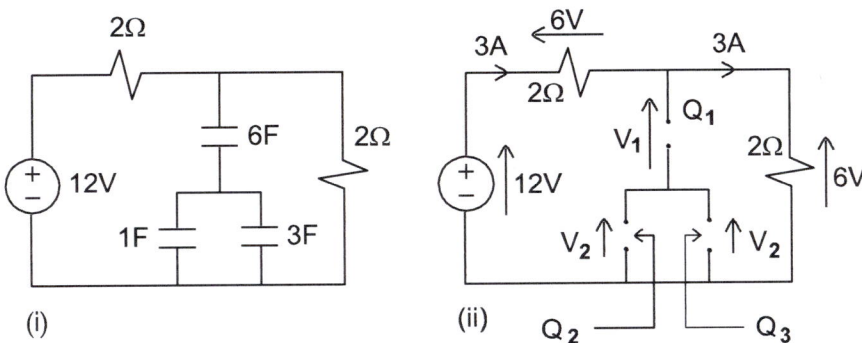

(i)   (ii)

Fig. 2.8(ii) shows the circuit of (i) under steady-state conditions.

When the circuit of Fig. 2.8(i) is given enough time for the capacitors to charge fully and for currents and voltages to reach their final steady-state (constant) values, the circuit can be represented with capacitors replaced with open circuits, as in Fig. 2.8(ii). This follows from $i=C(dv/dt)$ for the capacitor where, under steady-state conditions the voltage across the capacitor does not change with time and dv/dt , and the capacitor current, is zero. The capacitor can then be represented by an open circuit.

Only the outermost loop in Fig. 2.8 (ii) carries a current (of 3A) and, therefore, 6V appears across each of the two equal value resistors. The voltage across the 6F capacitor and the charge on it are V_1 and Q_1, respectively. The same voltage, V_2, appears across the other two capacitors that are connected in parallel. Q_2 and Q_3 are the charges on the 1F and 3F capacitors, respectively. $Q_2 \neq Q_3$ because the two capacitors in parallel have different capacitances.

The charge on the equivalent capacitance of the two capacitors in parallel is equal to $Q_2 + Q_3$. Also, since the equivalent capacitor is in series with the 6F one

$$Q_1 = Q_2 + Q_3 \tag{2.37}$$
$$6V_1 = 1V_2 + 3V_2 \tag{2.38}$$
$$3V_1 = 2V_2$$

Also, from KVL applied to the loop on the right hand side of Fig. 2.8 (ii),

$$V_1 + V_2 = 6 \tag{2.39}$$

Solving Equs (2.38) and (2.39) simultaneously gives

$$V_1 = 2.4 \text{ V} \tag{2.40}$$
$$V_2 = 3.6 \text{ V} \tag{2.41}$$

And since $Q=CV$

$$Q_1 = 6V_1 = 6 \times 2.4 = 14.4 \text{ C} \tag{2.42}$$
$$Q_2 = 1V_2 = 3.6 \text{ C} \tag{2.43}$$
$$Q_3 = 3V_2 = 3 \times 3.6 = 10.8 \text{ C} \tag{2.44}$$

(An alternative method of solving this problem is to start by combining the two parallel capacitors into an equivalent capacitor of capacitance 4F. The charge on this equivalent capacitor equals the charge on the 6F capacitor, Q_1, since the two are in series with one another. From $Q=CV$ and KVL, $(Q_1/4) + (Q_1/6) = 6$ which gives $Q_1 = 14.4$ C. Using $Q=CV$, again, gives $V_1 = 14.4/6 = 2.4$ V and $V_2 = 14.4/4 = 3.6$ V. Q_2 and Q_3 then follow as in Equs. 2.43 and 2.44 above).

2.4 Inductor current, voltage and energy relationships
2.4.1 Current -voltage relationship

A coiled wire constitutes an inductor. An inductor is a circuit element with a current-voltage relationship given by

$$v(t) = L\frac{di(t)}{dt} \tag{2.45}$$

where L is the inductance of the element in henrys (abbreviated H). Fig. 2.9 is the schematic representation of the inductor, showing the relative directions of the current through the element and the voltage drop across it. Equation (2.45) states that the voltage drop across the inductor is proportional to the time rate of change of the current through it.

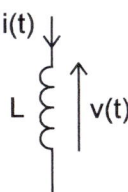

Fig. 2.9 Schematic representation of inductor of inductance L henrys.

Equation 2.45 is arrived at by considering the magnetic field that exists around a current-carrying coil. The current i flowing through a conducting wire with a single loop produces a magnetic field around the wire so that the magnetic flux ψ linking, or passing through, the loop is proportional to the current. Then

$$\psi = Li \tag{2.46}$$

where the inductance L is the constant of proportionality. Since ψ is measured in units of webers (abbreviated Wb) the units of

75

inductance, henrys, are equivalent to webers/ampere (Wb/A) or from Equ. (2.45) to (volts seconds)/ampere.

For a coil of N turns, the magnetic flux passing through the coil is $N\psi$, and Equ. (2.46) becomes

$$N\psi = Li \qquad (2.47)$$

Faraday's law states that the voltage induced in an N-turn coil when the flux ψ passing through it changes at the rate of $d\psi/dt$ is

$$v(t) = N\frac{d\psi(t)}{dt} \qquad (2.48)$$

Then Equs. 2.47 and 2.48 give

$$v(t) = N\frac{d\psi(t)}{dt} = L\frac{di(t)}{dt} \qquad (2.49)$$

which is Equ. (2.45) and relates the induced voltage to the rate of change of current. Under dc conditions, the current and magnetic flux are constant and the induced voltage is zero. An increase or decay in current with time results in corresponding changes in magnetic flux and induces a voltage. According to Lenz's law, the direction of the induced voltage is such that it produces a current that opposes the flux change. This in effect means that the induced voltage opposes sudden changes in the current through the inductor, so that the change in current is not instantaneous but gradual.

The formula for the current through the inductor at a general time t is obtained by integrating Equ (2.49) to get

$$i = \frac{1}{L}\int_0^t v(t)dt \qquad (2.50)$$

If an initial current flows through the inductor at $t=0^-$, the value given by Equ. (2.50) is added to the initial current to find the current at t.

2.4.2 Energy stored in inductor

The power delivered to the inductor at time t is given by

$$p(t) = v(t)i(t) \qquad (2.51)$$

Substituting for the voltage from Equ. (2.45) into (2.51) gives

$$p(t) = Li(t)\frac{di(t)}{dt} \qquad (2.51)$$

The energy stored in the magnetic field of the inductor is then

$$U = \int_0^t p(t)dt \qquad (2.52)$$

$$U = \int_0^t Li(t)\frac{di}{dt}dt \qquad (2.53)$$

$$U = \int_{i(0)}^{i(t)} Li(t)di \qquad (2.54)$$

$$U = \frac{1}{2}Li^2(t) - \frac{1}{2}Li^2(0) \qquad (2.55)$$

where $i(0)$ is the initial current through the inductor and $[Li^2(0)]/2$ is the initial energy stored in the inductor.

In general, under steady-state conditions, if the current through the inductor, of inductance L, is a constant, I, then the energy stored in the inductor depends on the current and is given by

$$U = \frac{1}{2}LI^2 \qquad (2.56)$$

Since the energy of a system cannot change instantaneously, then similarly, the current through the inductor cannot change instantaneously.

2.4.3 Inductance of a coil

The inductance of a single layer coil is approximately equal to

$$L = \frac{\mu N^2 A}{l} \quad \text{Henry} \qquad (2.56)\text{a}$$

where μ is the permeability of the core on which the coil is wound ($\mu = \mu_o \mu_r$, where $\mu_o = 4\pi \times 10^{-7}$ Henry/meter is the permeability of free space, and μ_r the permeability of the core material, see Section 5.2), N is the number of turns, A is the area of the coil (area of a loop) in square meters, and l is the coil length in meters.

Example 2.7: Initial and final conditions in RL circuit

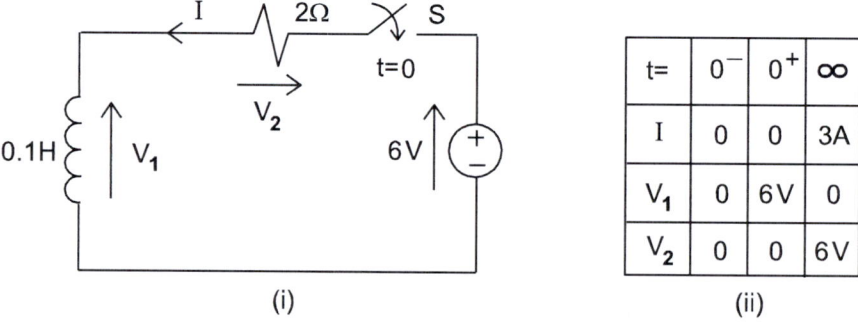

t=	0^-	0^+	∞
I	0	0	3A
V_1	0	6V	0
V_2	0	0	6V

(i)   (ii)

Fig. 2.10 Find the initial and final values of V_1, V_2 and I (Example 2.7).

78

In the circuit of Fig. 2.10(i) the switch S is open for time $t<0$. At $t = 0$ the switch is closed. Find V_1, V_2 and I immediately before the switch is closed at $t = 0^-$, immediately after the switch is closed at $t = 0^+$, and when the switch has been closed for a long time $t = \infty$, to complete the table in Fig. 2.10 (ii).

No current flows through the inductor for $t<0$ therefore $I=0$ at $t=0^-$. At $t=0^+$ $I=0$ because the current through the inductor cannot change instantaneously and so it must be the same immediately before and immediately after the switch is closed. Then, from Ohm's law, the voltage across the resistor at $t=0^+$, V_2, is zero, and from KVL the supply voltage of 6V appears across the inductor. As time increases the current through the circuit increases until the voltage across the resistor is equal and opposite to the supply voltage when the current reaches its maximum value. At this point the circuit is in the steady-state condition and the constant current that flows is $(V_2/2) = 6/2 = 3$A.

Under steady-state conditions di/dt is zero, and since the voltage across the inductor is proportional to di/dt, the voltage across the inductor is zero. The inductor is then represented by a short circuit.

When the switch is initially closed, the voltage across the inductor changes from 0 to 6V instantaneously, therefore, the rate of change of current at $t=0^+$ is $(V_1/L) = (6/0.1) = 60$ amperes per second. At $t=0^+$ the current through the inductor is zero but the rate at which the current is rising is at its maximum. At $t = \infty$ the current through the inductor is at its maximum but its rate of increase with time is zero. An inductor is analogous to a flywheel in an engine; the sudden application of force can cause a sudden large acceleration but inertia prevents the engine from undergoing a sudden change in angular velocity and the velocity, which is analogous to the inductor current, increases gradually.

2.5 Combining inductors in series and parallel
2.5.1 Series combination

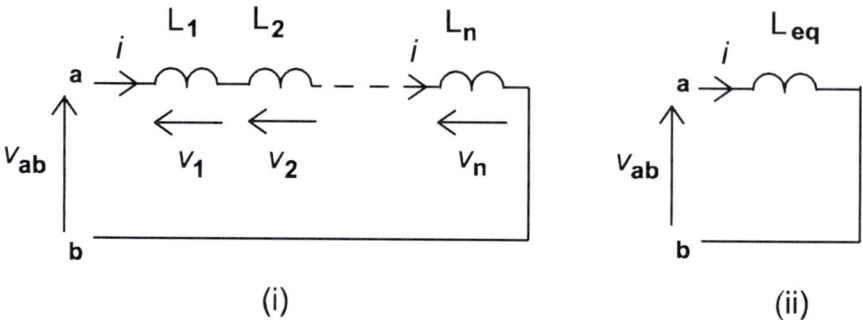

Fig. 2.11 Combining inductors in series.

Fig. 2.11(i) shows inductors connected in series, between terminals a and b, and the aim is to find how the inductors can be combined so that one equivalent inductor L_{eq} replaces them. In Fig 2.11(i) the current through each inductor is the same and KVL gives

$$v_{ab} = v_1 + v_2 + ... + v_n \tag{2.57}$$

$$v_{ab} = L_1 \frac{di}{dt} + L_2 \frac{di}{dt} + ... + L_n \frac{di}{dt} \tag{2.58}$$

$$v_{ab} = (L_1 + L_2 + ... + L_n) \frac{di}{dt} \tag{2.59}$$

$$v_{ab} = L_{eq} \frac{di}{dt} \tag{2.60}$$

Therefore, the inductors in Fig. 2.11(i) can be replaced by their equivalent, L_{eq}, shown in Fig. 2.11(ii), where

$$L_{eq} = L_1 + L_2 + ... + L_n \tag{2.61}$$

This shows that inductors in series combine as resistors in series.

2.5.2 Parallel combination

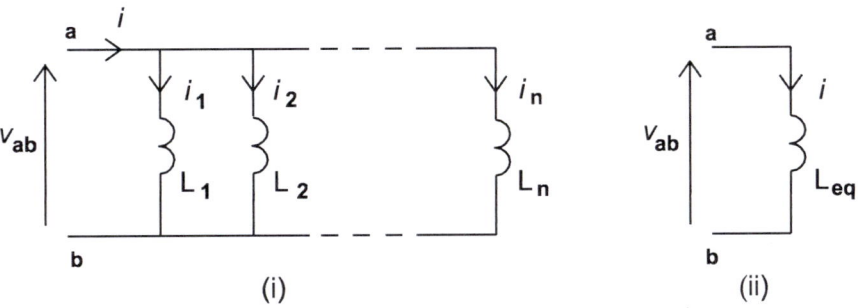

Fig. 2.12 Combining inductors connected in parallel.

Fig. 2.12(i) shows inductors connected in parallel and Fig 2.12(ii) shows the inductors replaced by their equivalent inductor L_{eq}. The voltage across each inductor in Fig. 2.12(i) is the same. KCL gives that at time t,

$$i = i_1 + i_2 + ... + i_n \tag{2.62}$$

$$i = \frac{1}{L_1} \int_0^t v\,dt + \frac{1}{L_2} \int_0^t v\,dt + ... + \frac{1}{L_n} \int_0^t v\,dt \tag{2.63}$$

$$i = (\frac{1}{L_1} + \frac{1}{L_2} + ... + \frac{1}{L_n}) \int_0^t v\,dt \tag{2.64}$$

$$i = \frac{1}{L_{eq}} \int_0^t v\,dt \tag{2.65}$$

Therefore, the equivalent inductor is given by

$$L_{eq} = \frac{1}{\left(\dfrac{1}{L_1} + \dfrac{1}{L_2} + ... + \dfrac{1}{L_n} \right)} \tag{2.66}$$

81

Example 2.8: Initial and final conditions

In the circuit of Fig. 2.13(i) the switch S is open for time $t<0$ and the circuit is in the steady-state condition (constant currents and voltages). At $t = 0$ the switch is closed. Find i_1, i_2, i_3, v_C and v_L immediately before the switch is closed at $t = 0^-$, immediately after the switch is closed at $t = 0^+$, and when the switch has been closed for a long time $t = \infty$, to complete the table in Fig. 2.13 (ii).

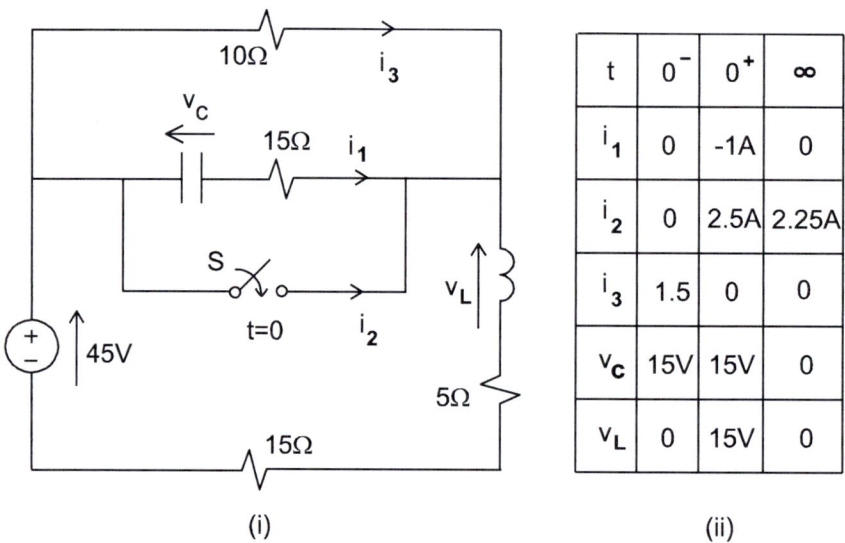

t	0^-	0^+	∞
i_1	0	-1A	0
i_2	0	2.5A	2.25A
i_3	1.5	0	0
v_C	15V	15V	0
v_L	0	15V	0

(i)   (ii)

Fig. 2.13 Circuit and table for Example 2.8.

82

a) $t=0^-$ The circuit for $t<0$, up to $t = 0^-$, appears in Fig. 2.14. The capacitor is replaced with an open and the inductor with a short

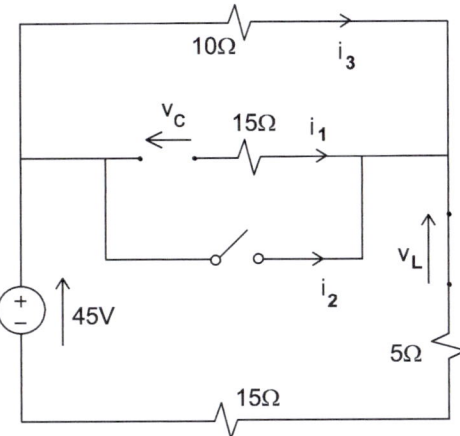

Fig. 2.14 Circuit of Fig. 2.13 at $t = 0^-$ (Example 2.8).

circuit. Currents i_1 and i_2 are both zero. Current i_3 flows in the outermost loop (that includes the inductor), and is given by Ohm's law, or KVL, as

$$i_3 = 45/(10+5+15) = 1.5 \text{ A} \qquad (2.67)$$

Since $i_1 = 0$, the voltage across the 15Ω resistor to the right of the capacitor, is, from Ohm's law, zero. KVL applied to the top loop then shows that the voltage across the capacitor is the same as the voltage across the 10Ω resistor and is given by

$$v_C = 10 \times i_3 = 10 \times 1.5 = 15 \text{ V} \qquad (2.68)$$

v_L is the voltage across a short circuit which is zero.

b) $t=0^+$ The circuit for $t = 0^+$ appears in Fig. 2.15.

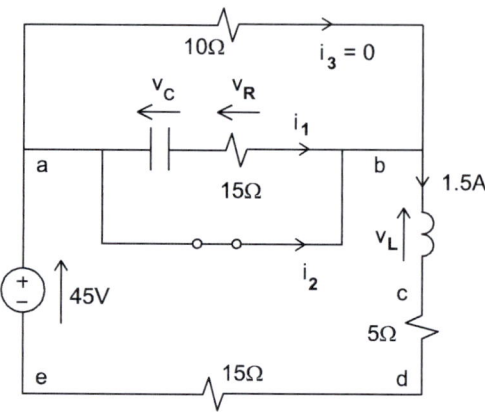

Fig. 2.15 Circuit of Fig. 2.13 at $t = 0^+$ (Example 2.8).

The voltage across the capacitor and the current through the inductor cannot change instantaneously, so they stay at 15V and 1.5A, as at $t = 0^-$. KVL applied to the inner loop containing the capacitor and the 15Ω resistor gives

$$v_C + v_R = 0 \qquad (2.69)$$
$$15 + 15i_1 = 0 \qquad (2.70)$$
$$i_1 = -1\text{A} \qquad (2.71)$$

Nodes a and b are shorted together through the closed switch. Since these nodes are also the terminals of the 10Ω resistor, the voltage across this resistor is zero, and from Ohm's law, the current through it, i_3, is also zero. Then, since the inductor current is 1.5A, KCL at node b gives

$$i_1 + i_2 = 1.5 \qquad (2.72)$$
$$-1 + i_2 = 1.5 \qquad (2.73)$$
$$i_2 = 2.5\,\text{A} \qquad (2.74)$$

Since the voltage between nodes a and b is zero, applying KVL in loop *abcde gives* the voltage across the inductor

$$45 - v_L - 1.5 \times 5 - 1.5 \times 15 = 0 \tag{2.75}$$
$$v_L = 15 \text{ V}. \tag{2.76}$$

c) $t = \infty$. The circuit for $t = \infty$ appears in Fig. 2.16. Because this is a steady state situation, the capacitor is replaced with an open and the inductor with a short circuit.

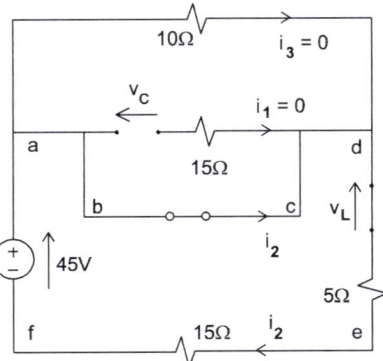

Fig. 2.16. Circuit of Fig. 2.13 at $t = \infty$ (Example 2.8).

The voltage across the inductor is zero. No current can flow through the open circuited capacitor and therefore, i_1 is zero. Furthermore, since the capacitor and the 15Ω resistor on its right are short circuited by the closed switch, the voltage across the capacitor v_C is zero. The switch shorts out the 10Ω resistor and therefore, i_3 is also zero. The loop *abcdef* is the only active loop carrying the current i_2. KVL in this loop gives

$$i_2 = (45/20) = 2.25 \text{ A} \tag{2.77}$$

85

Example 2.9: Initial and final conditions

The switch S in the circuit in Fig. 2.17 is open for $t<0$ and the circuit is in the steady-state condition (constant currents and voltages) for that time. The switch is closed at $t=0$. Find i_1, i_2, i_3 and v_C immediately before, and immediately after, $t=0$, and when the circuit has stabilized once again at $t=\infty$. Assume that the current directions and voltage polarities are as shown in Fig. 2.17(i).

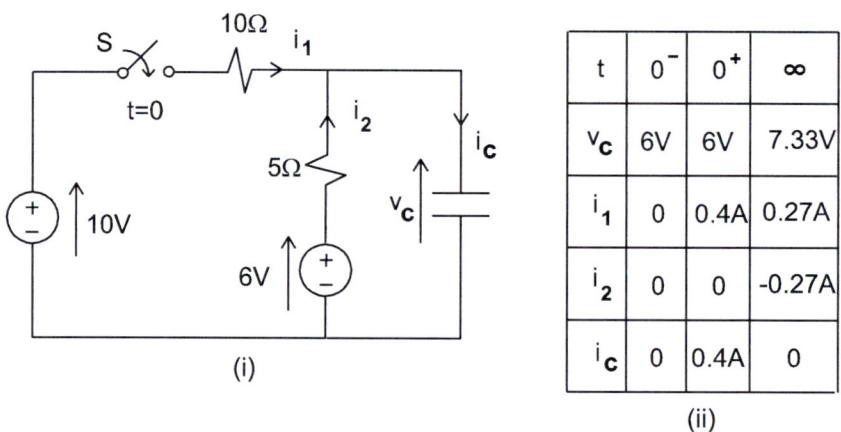

t	0^-	0^+	∞
v_C	6V	6V	7.33V
i_1	0	0.4A	0.27A
i_2	0	0	-0.27A
i_C	0	0.4A	0

(i)   (ii)

Fig. 2.17 Circuit and table for Example 2.9.

a) $t=0^-$ At $t<0$, under steady-state conditions, the capacitor is an open circuit and $i_1 = i_2 = i_c = 0$. KVL for the loop on the right hand side gives

$$6 + 0 - v_C = 0 \tag{2.78}$$

$$v_C = 6 \text{ V} \tag{2.79}$$

b) t=0⁺   The voltage across the capacitor cannot change instantaneously and therefore, immediately after closing the switch

$$v_C(0^+) = v_C(0^-) = 6 \text{ V} \tag{2.80}$$

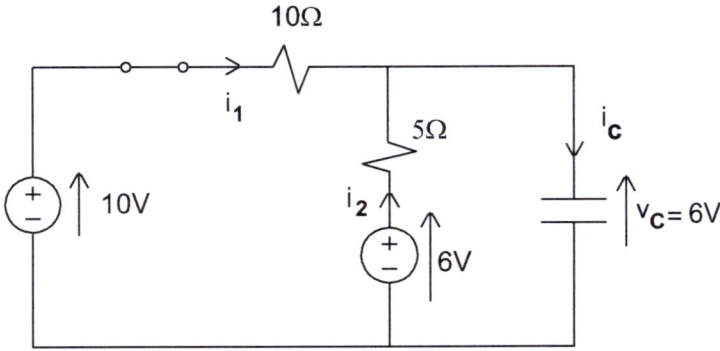

Fig. 2.18 The circuit of Fig. 2.17 at $t=0^+$ (Example 2.9).

Then, from KVL, the voltage across the 5Ω resistor is zero and therefore, from Ohm's law, $i_2=0$. The voltage across the 10Ω resistor is (10-6) V and therefore, once again from Ohm's law

$$i_1 = \frac{(10-6)}{10} = 0.4 \text{ A} \tag{2.81}$$

From KCL

$$i_C = i_1 + i_2 = 0.4 \text{ A} \tag{2.82}$$

c) t=∞

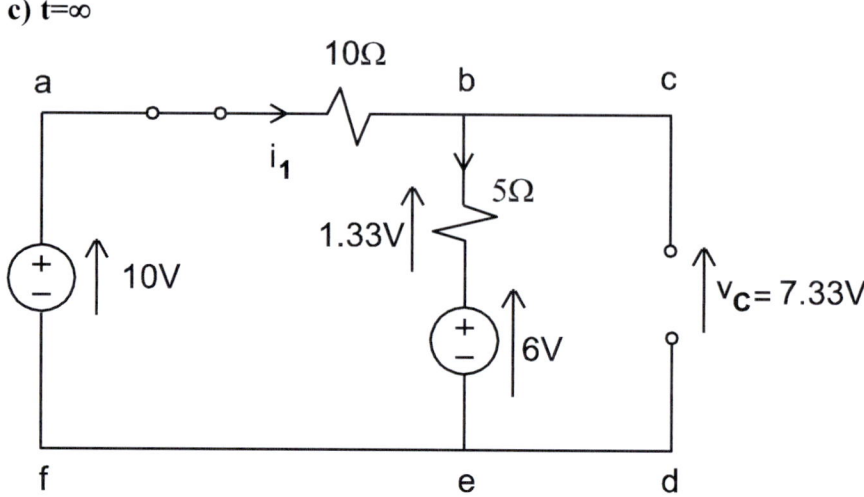

Fig. 2.19 The circuit of Fig. 2.17 at *t*=∞ (Example 2.9).

When the circuit stabilizes at *t*=∞, the circuit of Fig. 2.19 applies. The capacitor is an open circuit and i_C=0. Considering loop *abef* (10-6) V appears across the series combination of the 10Ω and 5Ω resistors and therefore, Ohm's law gives

$$i_1 = \frac{(10-6)}{(10+5)} = 0.27 \text{ A} \tag{2.83}$$

Ohm's law gives the voltage across the 5Ω resistor as (0.27×5) =1.33V. KVL in loop *bcde* then gives

$$6 + 1.33 - v_C = 0 \tag{2.84}$$
$$v_C = 7.33 \text{ V} \tag{2.85}$$

Chapter 3
Methods of ac circuit analysis

3.1 Introduction

In Chapter 1 electrical laws and methods were applied to the analysis of resistive circuits powered by dc (i.e. constant) sources. In the present chapter the same principles will be applied to circuits involving resistors, capacitors and inductors powered by ac (alternating) sources of electricity. These are voltage and current sources that provide periodic electrical waveforms at their output terminals. These are sinusoidal sources which result in sinusoidally varying voltages and currents in different branches of the circuit. Sinusoidal sources are represented as phasors, and voltages and currents they produce in different parts of the circuit are calculated using phasor algebra. Phasor algebra will also be used in Chapter 4 to calculate power flow in ac circuits.

The number of times an ac waveform repeats itself every second is the frequency of the signal. Frequency is measured in hertz (Hz). Electrical power is transmitted at 60Hz (in the US) and 50Hz (in Europe). An ac voltage can be stepped up or down using a transformer. The transformer will be discussed further in Chapter 5. A use of this device is to step-up the amplitude of ac voltage from its generation value to a large enough magnitude to enable the transmission of electrical power over long distances. The voltage is subsequently stepped down at its destination using a step-down transformer. A transformer works with ac not dc; this gives ac energy an advantage over dc.

3.2 Sinusoidal voltage and current

Sinusoidal voltage and current sources produce periodic electrical waveforms. A periodic waveform repeats itself every T seconds to display a cycle. T is the period of the function representing the waveform. The number of cycles displayed in one second is the cyclic frequency f of the waveform. Cyclic frequency is measured in cycles per second, or hertz (Hz), where

$$f = \frac{1}{T} \qquad (3.1)$$

If $f = 60$ Hz then the signal completes 60 cycles every second.

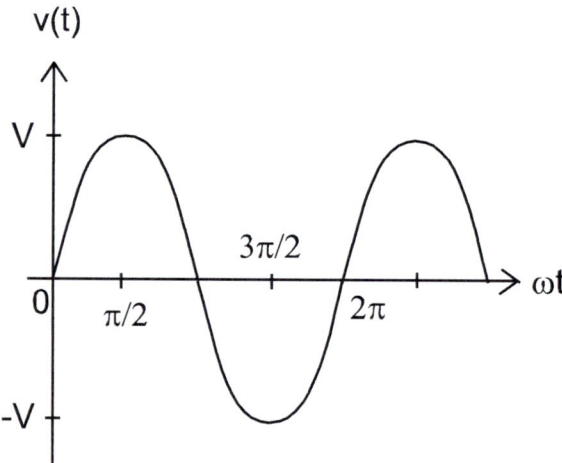

Figure 3.1 Plot of $v(t)=V\sin(\omega t)$.

Figure 3.1 shows the plot of the sine function

$$v(t) = V \sin(\omega t) \qquad (3.2)$$

representing the variation with time of the voltage across the terminals of a sinusoidal source. The horizontal axis is proportional to time t seconds and is measured in terms of ωt, where ω is the angular or radian frequency, measured in radians per second and, therefore, ωt is measured in radians. (ω is also called the angular velocity). The angular frequency is given by

$$\omega = 2\pi f \tag{3.3}.$$

The waveform starts at the origin and reaches its maximum value, or amplitude, V at $\omega t = \pi/2$. The function returns to zero at $\omega t = \pi$ and then turns negative, reaching a minimum of $-V$ at $\omega t = 3\pi/2$, before returning to zero at $\omega t = 2\pi$. At

$$\omega t = 2\pi \tag{3.4}$$

$$\frac{2\pi t}{T} = 2\pi \tag{3.5}$$

$$t = T \tag{3.6}.$$

Since the time taken by a periodic function to complete one cycle

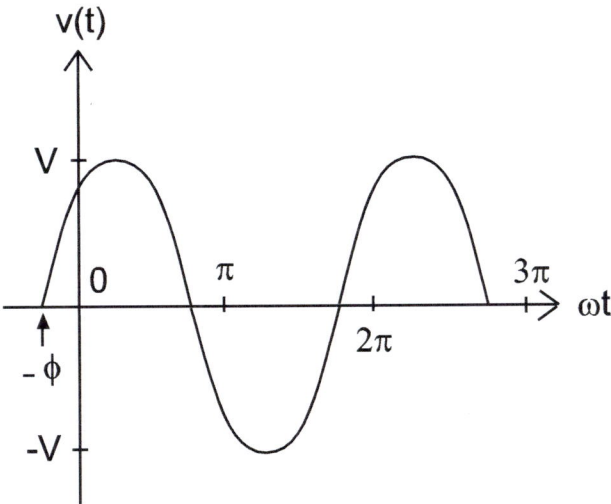

Figure 3.2 Plot of $v(t) = V\sin(\omega t + \phi)$.

91

is the period T, then at $\omega t = 2\pi$ radians, the sine wave has gone through a complete cycle. The function is periodic and therefore repeats itself every T seconds. The sine function is also an alternating waveform since it continually changes between a positive and a negative peak.

Adding a phase angle ϕ to the argument of the sine function of Equ. (3.2), shifts the function in the negative ωt direction by that phase angle, i.e. advances the function by ϕ (see Fig. 3.2). The function sin ($\omega t + \phi$) starts ϕ radians earlier than sin (ωt); the function sin ($\omega t + \phi$) leads the function sin (ωt) by ϕ radians.

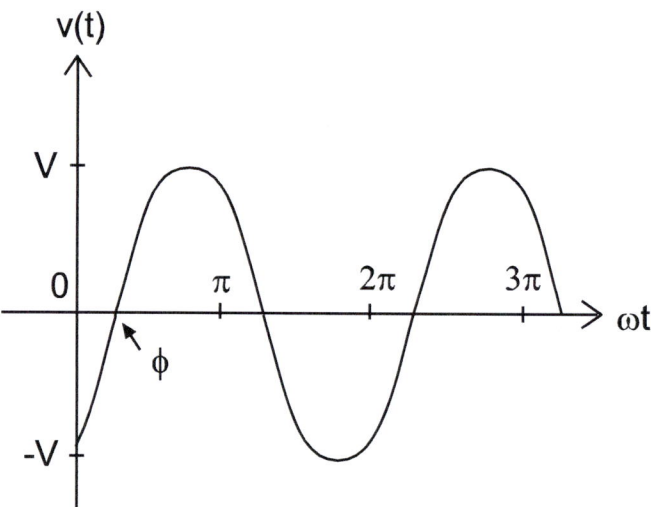

Figure 3.3 Plot of $v(t) = V \sin(\omega t - \phi)$.

Subtracting a phase angle ϕ from the argument of the sine function of Equ. (3.2), shifts the function in the positive ωt direction by that phase angle, i.e. delays the function by ϕ (see Fig. 3.3). The function sin ($\omega t - \phi$) starts ϕ radians later than sin (ωt); the function sin ($\omega t - \phi$) lags the function sin (ωt) by ϕ radians.

Figure 3.4 shows the cosine function $v(t)=V\cos(\omega t)$. It is the same as the sine function but advanced in time by $\pi/2$ so that the cosine function displays its first peak at $\omega t=0$, compared to the first peak of the sine function that occurs at $\pi/2$.

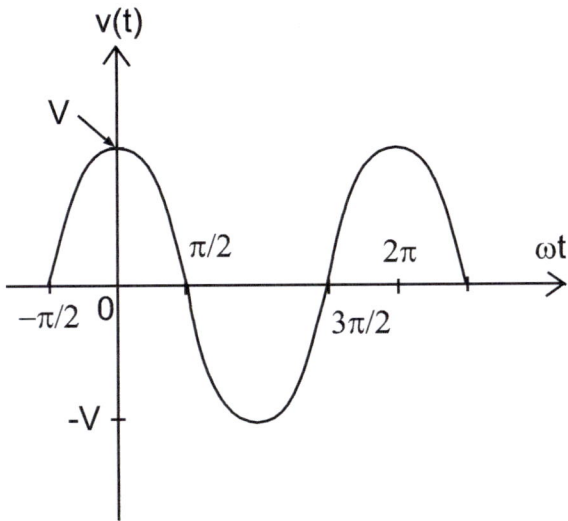

Figure 3.4 Plot of $v(t)=V\cos(\omega t)$.

The conversion between the two functions is done according to

$$V\cos(\omega t) = V\sin(\omega t + \frac{\pi}{2})$$ (3.7)

$$= V\sin(\omega t + 90°)$$ (3.8)

$$V\sin(\omega t) = V\cos(\omega t - \frac{\pi}{2})$$ (3.9)

$$= V\cos(\omega t - 90°)$$ (3.10)

3.3 RMS value of sinusoidal quantity

The root-mean-square (rms) value of an alternating current is that current which will produce the same amount of heat in a resistor as an equivalent direct current. For instance, if the rms current of an alternating current flowing through a resistor of $R\Omega$ is I_{rms}, then the power consumed by the resistor (referred to as the average power) is equal to the power consumed if a direct current I_{dc}, equal in value to I_{rms}, were to flow through the resistor. This power is

$$I_{rms}^2 R = I_{dc}^2 R \qquad (3.11)$$

The rms value of a function $i(t)$ is the square root of the average of $i^2(t)$ and is given by

$$I_{rms} = \left[\frac{1}{T} \int_0^T i^2(t) dt \right]^{1/2} \qquad (3.12).$$

The rms value of a sinusoidal current $i(t) = I \sin(\omega t)$ is

$$I_{rms} = \left[\frac{1}{T} \int_0^T I^2 \sin^2(\frac{2\pi t}{T}) dt \right]^{1/2} \qquad (3.13)$$

$$I_{rms} = \frac{I}{\sqrt{2}} \qquad (3.14)$$

$$I_{rms} = 0.707 I \qquad (3.15)$$

Therefore, dividing the amplitude, of a sinusoidally varying current or voltage, gives the rms value of the respective variable. Rms values are used in calculations of power in systems supplied by sinusoidal sources.

3.4 Phasor representation of sinusoids

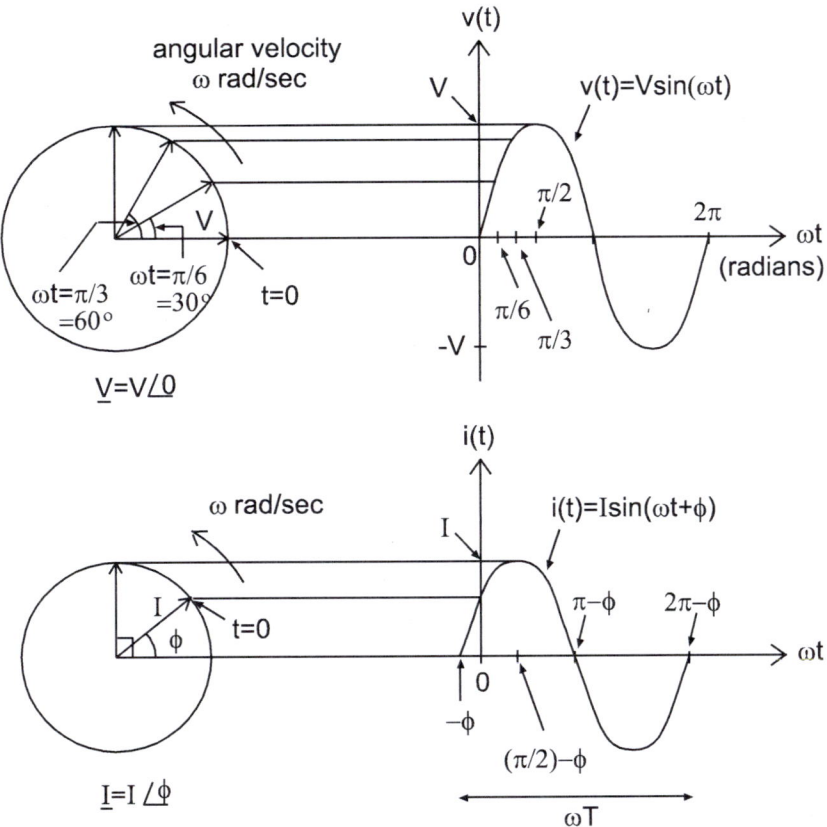

Fig. 3.5 The vectors $\underline{V}$ and $\underline{I}$ rotate in an anticlockwise direction, with an angular velocity of ω radians per second about the center of the circle, and their respective projections on the vertical axis, as a function of time, give their corresponding sinusoidal waveforms.

Figure 3.5 shows how a rotating vector, or phasor, on the left generates a sine wave on the right. Assume that a voltage $v(t)=V\sin(\omega t)$, which is applied across a branch in a circuit, results in a current $i(t)=I\sin(\omega t+\phi)$ through the branch, and that these two waveforms need to be represented as phasors. Then the vector, of magnitude V, aligned with the horizontal at time $t=0$, and rotating in an anticlockwise direction with an angular velocity of ω radians per second, will generate $v(t)=V\sin(\omega t)$, if the projection of this vector along the vertical axis is plotted as a function of ωt. Similarly, a vector of magnitude I, making an angle of ϕ with the horizontal at time $t=0$, and rotating in an anticlockwise direction with an angular velocity of ω radians per second, will generate the function $i(t)=I\sin(\omega t+\phi)$, if the projection of this vector along the vertical axis is plotted as a function of ωt. As a result, the projection of the voltage phasor along the vertical is zero at $t=0$ and the corresponding sine wave starts at the origin, whereas at ωt $=\pi/2$ the phasor is at 90° to the horizontal, with its projection along the vertical at its maximum, at which time the sine wave displays its peak voltage or amplitude.

The sinusoidal variation of a voltage can be measured and displayed on an oscilloscope screen. A phasor, however, is a mathematical description of a sinusoid, which is used in calculations to predict the effect of sinusoidal sources on different parts of the ac circuit. For these calculations, the phasor can be represented in three ways.

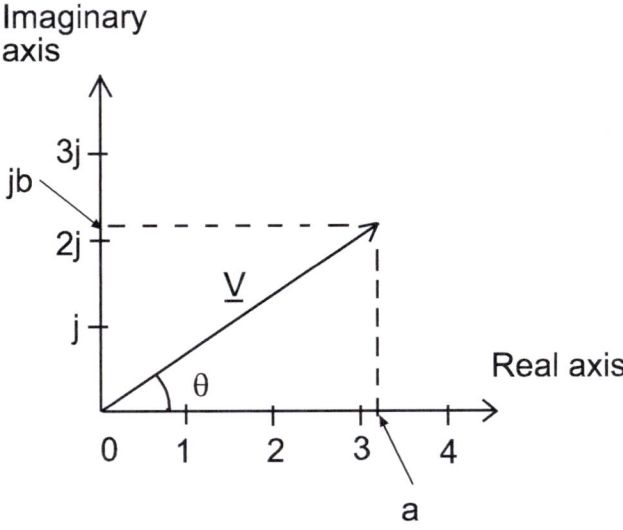

Fig. 3.6 Representation of phasor $\underline{V}$ in the complex plane. The length, or magnitude, of the phasor is V and its phase angle is θ. The magnitude of the components of the phasor along the real and imaginary axes are a and b, respectively.

Firstly, with the center of the circle about which the phasor rotates as the origin, a phasor $\underline{V}$ is described in the rectangular form in terms of its components along the vertical and horizontal axes at $t=0$, where the horizontal axis is the real axis and the vertical axis the imaginary one (see Fig. 3.6). $\underline{V}$ is a complex quantity given by

$$\underline{V} = a + jb \qquad (3.16)$$
$$a = V\cos\theta = \mathrm{Re}(\underline{V}) \qquad (3.17)$$
$$b = V\sin\theta = \mathrm{Im}(\underline{V}) \qquad (3.18)$$
$$\theta = \tan^{-1}\left(\frac{b}{a}\right) \qquad (3.19)$$

where $j=\sqrt{-1}$ and $j^2=-1$. $\mathrm{Re}(\underline{V})$ and $\mathrm{Im}(\underline{V})$ represent the real and imaginary part of $\underline{V}$, respectively, while θ is the phase angle of the phasor.

Secondly, if the magnitude of the phasor is V, and the angle it makes with the real axis at $t=0$ is θ, then in the polar representation the phasor is

$$\underline{V} = V\angle\theta \qquad (3.20)$$

Thirdly,

$$\begin{aligned} \underline{V} &= a + jb \\ &= V\cos\theta + jV\sin\theta \\ &= V(\cos\theta + j\sin\theta) \end{aligned} \qquad (3.21)$$

Euler's identity gives

$$e^{j\theta} = \cos\theta + j\sin\theta \qquad (3.22)$$

and substituting, from Equ. 3.22 into Equ. 3.21, gives the exponential polar form of phasor $\underline{V}$ of magnitude V and phase angle θ as

$$\underline{V} = Ve^{j\theta} \qquad (3.23)$$

A phasor of magnitude unity and phase angle θ is given in the exponential polar form as $e^{j\theta}$.

3.5 Phasor operations

It is easy to make calculation errors in phasor operations. The use of calculators that perform complex algebra is very helpful in reducing these errors. Although the capabilities of different calculators vary, in general some operations can be performed in fewer steps if the phasors are first expressed in a particular form. For instance, adding or subtracting phasors is easiest when they are in the rectangular form, so that for two phasors $\underline{V}_1$ and $\underline{V}_2$

$$\underline{V}_1 = a + jb \tag{3.24}$$

$$\underline{V}_2 = c + jd \tag{3.25}$$

$$\underline{V}_1 + \underline{V}_2 = (a+c) + j(b+d) \tag{3.26}$$

$$\underline{V}_1 - \underline{V}_2 = (a-c) + j(b-d) \tag{3.27}$$

Multiplication and division of phasors is easiest when they are in the polar form, so that for phasors $\underline{V}_1$ and $\underline{V}_2$

$$\underline{V}_1 = V_1 \angle \theta_1 \tag{3.28}$$

$$\underline{V}_2 = V_2 \angle \theta_2 \tag{3.29}$$

$$\underline{V}_1 \underline{V}_2 = V_1 V_2 \angle (\theta_1 + \theta_2) \tag{3.30}$$

$$\frac{\underline{V}_1}{\underline{V}_2} = \frac{V_1}{V_2} \angle (\theta_1 - \theta_2) \tag{3.31}$$

Differentiation of a phasor is best carried out with the phasor in the exponential polar form.

3.6 Effect of phasor source on a branch voltage and current

Let the circuit be excited by a voltage source that provides a voltage across its terminals that varies, with time, according to the function

$$V_s \sin(\omega t + \phi) \tag{3.32}$$

Then this voltage is the imaginary part of

$$V_s \left[\cos(\omega t + \phi) + j\sin(\omega t + \phi) \right] \tag{3.33}$$
$$= V_s e^{j(\omega t + \phi)} \tag{3.34}$$
$$= V_s e^{j\omega t} e^{j\phi} \tag{3.35}$$
$$= (V_s \angle \phi) e^{j\omega t} \tag{3.36}$$

The source voltage causes a voltage across a branch and a current through the branch, somewhere in the circuit, where these are

$$V \sin(\omega t + \theta_V) \tag{3.37}$$
$$= \text{Im}\{V \left[\cos(\omega t + \theta_V) + j\sin(\omega t + \theta_V) \right]\} \tag{3.38}$$
$$= \text{Im}\{(V \angle \theta_V) e^{j\omega t}\} \tag{3.39}$$

$$I \sin(\omega t + \theta_I) \tag{3.40}$$
$$= \text{Im}\{I \left[\cos(\omega t + \theta_I) + j\sin(\omega t + \theta_I) \right]\} \tag{3.41}$$
$$= \text{Im}\{(I \angle \theta_I) e^{j\omega t}\} \tag{3.42}$$

Therefore, the source, given by the phasor $V_s \angle \phi$, results in the branch voltage and current, given by the phasors $V \angle \theta_V$ and $I \angle \theta_I$ respectively, where all quantities vary as sine waves in this case, as dictated by the source, and with the same angular frequency ω of the source. The term $e^{j\omega t}$ is common to all voltages and currents in the circuit, and so it is left out, and what remains is a phasor represention of the circuit where the aim is, given $V_s\angle\phi$, to find $V\angle\theta_V$ and $I\angle\theta_I$.

3.7 Phasor representation of resistor, capacitor and inductor

The representation of the three circuit components in ac circuits is determined by their respective current-voltage characteristics.

3.7.1 Resistor

For the resistor, let an ac voltage across the resistor cause a current to flow through it so that the voltage and current are, respectively

$$v = Ve^{j(\omega t+\theta_V)} \tag{3.43}$$
$$i = Ie^{j(\omega t+\theta_I)} \tag{3.44}$$

Then Ohm's law gives

$$v = Ri \tag{3.45}$$
$$Ve^{j(\omega t+\theta_V)} = RIe^{j(\omega t+\theta_I)} \tag{3.46}$$
$$Ve^{j\omega t}e^{j\theta_V} = RIe^{j\omega t}e^{j\theta_I} \tag{3.47}$$
$$Ve^{j\theta_V} = RIe^{j\theta_I} \tag{3.48}$$
$$V\angle\theta_V = RI\angle\theta_I \tag{3.49}$$

$\underline{V} = V\angle\theta_V$ and $\underline{I} = I\angle\theta_I$ are the phasor voltage across, and the phasor current through, the resistor. Since the phase angles of the phasors on either side of Equ. 3.49 are equal, then $\theta_V = \theta_I$. This means that the phase angle of the voltage and of the current for the resistor are the same; the voltage and current are in phase. This is represented graphically in Fig. 3.7(i). The resistor is represented in a phasor circuit by its resistance of R ohms, in the same way as in a dc circuit.

3.7.2 Inductor

If the voltage and current for the inductor are v and i, as in Equs. 3.43 and 3.44, then the relationship between them is

$$v = L\frac{di}{dt} \tag{3.50}$$

$$Ve^{j(\omega t+\theta_V)} = L\frac{d}{dt}\left\{Ie^{j(\omega t+\theta_I)}\right\} \tag{3.51}$$

$$Ve^{j(\omega t+\theta_V)} = LIe^{j\theta_I}\frac{d}{dt}(e^{j\omega t}) \tag{3.52}$$

$$Ve^{j\theta_V}e^{j\omega t} = j\omega LIe^{j\theta_I}e^{j\omega t} \tag{3.53}$$

$$Ve^{j\theta_V} = j\omega LIe^{j\theta_I} \tag{3.54}$$

$$V\angle\theta_V = \{\omega L\angle(\pi/2)\}I\angle\theta_I \tag{3.55}$$

$$V\angle\theta_V = \omega LI\angle(\theta_I+\pi/2) \tag{3.56}$$

The step from Equ. (3.54) to Equ. (3.55) involved a rectangular to polar conversion $j\omega L = \omega L\angle(\pi/2)$. Equ. (3.56) shows that θ_V $=\theta_I+90°$. This means that the voltage phasor across the inductor $\underline{V} = V\angle\theta_V$ leads the current through the inductor $\underline{I} = I\angle\theta_I$ by $90°$, as shown in Fig. 3.7(ii).

Equ. 3.54 is analogous to Ohm's law and can be written as

$$\underline{V} = j\omega L\underline{I} \tag{3.57}$$

$$\underline{V} = \underline{Z}\underline{I} \tag{3.58}$$

where

$$\underline{Z} = j\omega L \tag{3.59}$$

$\underline{Z}$ is the complex resistance, or impedance, of the inductor, of inductance L, excited by a source that is varying at ω radians per second. At low frequencies the impedance of the inductor is small.

At dc the impedance is zero and the inductor is a short circuit. As the frequency increases, the impedance of the inductor increases and the current through it decreases. The inductor is represented in a phasor circuit by its impedance $j\omega L$. With ω in units of radians per second and L in henrys, $j\omega L$ is in ohms.

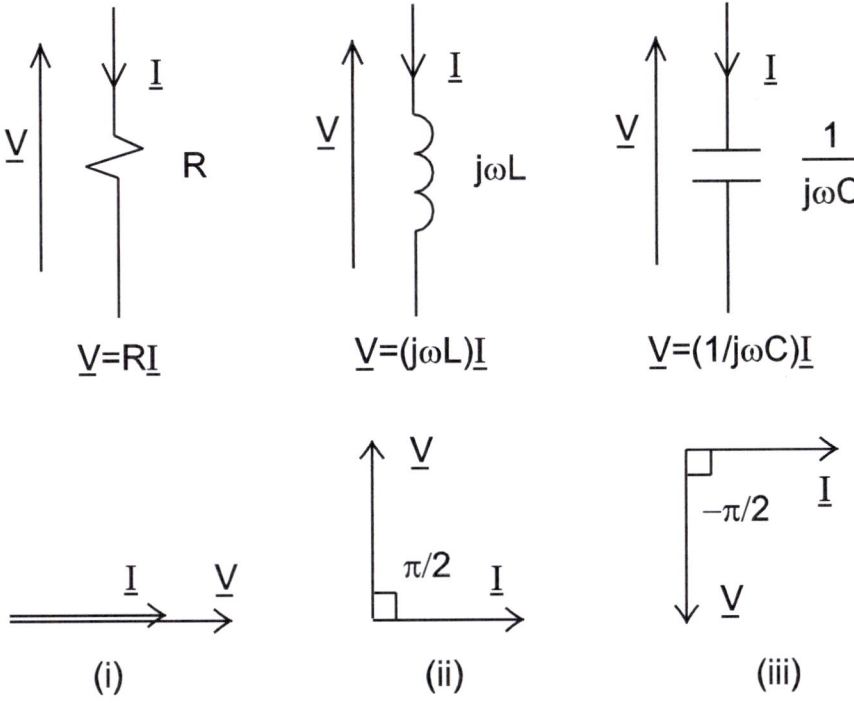

Fig. 3.7 In an ac circuit the resistor is represented by its resistance R, the inductor by its impedance $\underline{Z} = j\omega L$ and the capacitor by its impedance $\underline{Z} = (1/j\omega C)$, where $\omega = 2\pi f$ and the impedances are frequency dependent. The voltage across each element is then given by Ohm's law in the form $\underline{V} = \underline{Z}\underline{I}$ where $\underline{Z} = R$ for the resistor or the appropriate impedance for the inductor or capacitor.

3.7.3 Capacitor

Substituting for v and i from Equs. 3.43 and 3.44 into the current-voltage relationship for the capacitor gives

$$i = C\frac{dv}{dt} \tag{3.60}$$

$$Ie^{j(\omega t+\theta_I)} = C\frac{d}{dt}\left\{Ve^{j(\omega t+\theta_V)}\right\} \tag{3.61}$$

$$Ie^{j(\omega t+\theta_I)} = CVe^{j\theta_V}\frac{d}{dt}(e^{j\omega t}) \tag{3.62}$$

$$Ie^{j\theta_I}e^{j\omega t} = j\omega CVe^{j\theta_V}e^{j\omega t} \tag{3.63}$$

$$Ve^{j\theta_V} = (\frac{1}{j\omega C})Ie^{j\theta_I} \tag{3.64}$$

$$V\angle\theta_V = (-\frac{j}{\omega C})I\angle\theta_I \tag{3.65}$$

$$V\angle\theta_V = \{\frac{1}{\omega C}\angle(-\pi/2)\}(I\angle\theta_I) \tag{3.66}$$

$$V\angle\theta_V = \frac{I}{\omega C}\angle(\theta_I - \pi/2) \tag{3.67}.$$

Equ. 3.67 shows that the voltage across the capacitor lags the current through the capacitor by 90°. Equ. 3.64 is analogous to Ohm's law and can be written as

$$\underline{V} = \underline{Z}I \tag{3.68}$$

$$\underline{Z} = \frac{1}{j\omega C} \tag{3.69}$$

where $\underline{Z}$ is the complex resistance, or impedance, of the capacitor of capacitance C farads , excited by a source that is varying at ω radians per second. $\underline{Z}$ has the units of ohms. At low frequencies the impedance of the capacitor is high. At dc the capacitor has infinity impedance and is represented as an open circuit. As the frequency increases, the capacitor impedance decreases and the current through it increases. The capacitor is represented in a phasor circuit by its impedance $(1/j\omega C)$.

3.8 Phasor representation of sinusoidal energy sources

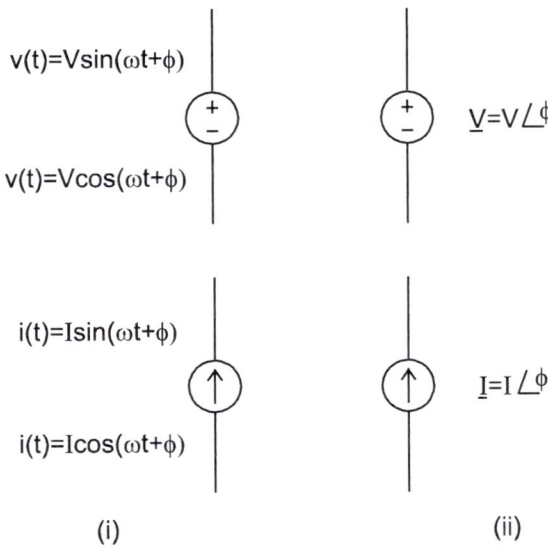

(i)   (ii)

Fig. 3.8 (i) Time domain and (ii) phasor representation of sinusoidal voltage and current sources.

Sinusoidally varying voltage and current sources are represented in the phasor circuit as phasors which show the amplitude and phase angle of the source, but do not specify whether the variation is a sine or a cosine one (see Fig. 3.8). A subsequent phasor calculation of an unknown circuit current or voltage will yield a phasor. If this phasor is then transformed into the time domain, it will vary as a sine or cosine function depending on the variation of the source.

The polarity of an ac voltage source and the direction of current for an ac current source that appear on the symbols representing them as in Fig. 3.8, apply for the first cycle of the waveform that they generate. Consequently, all current directions and voltage polarities in a circuit that the sources appear in, are then calculated with respect to the first half of a cycle. Voltage polarities and current directions for voltage and current sources are reversed in the second half of the cycle and so are all the directions of voltages and currents for the different branches of the circuit. Phasor calculations refer to electrical conditions in the first half of a cycle.

Example 3.1: Phasor calculation in single loop circuit

Find $i(t)$ in Fig.3.9(i).

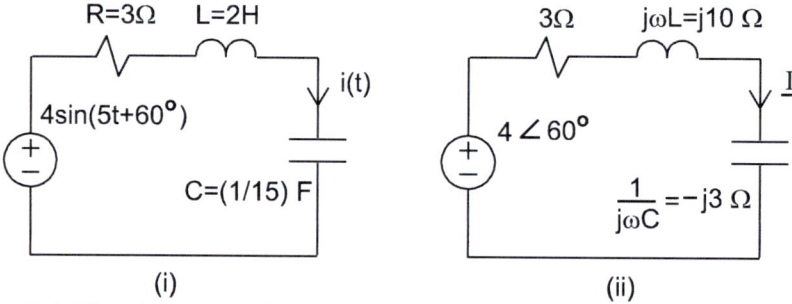

(i)   (ii)

Fig. 3.9 The time domain circuit in (i) and its phasor representation in (ii). Example 3.1.

Fig. 3.9(ii) shows the phasor representation of Fig. 3.9(i) using $\omega=5$ rad/sec. The voltage source sees a series combination of the three components $\underline{Z} = (3 + j10 - j3)$. Ohm's law then gives

$$\underline{I} = \frac{4\angle 60°}{(3 + j10 - j3)} = \frac{4\angle 60°}{(3 + j7)} = \frac{4\angle 60°}{7.62\angle 66.80°} = 0.53\angle -6.80° \text{ A}$$

$$i(t) = 0.53\sin(5t - 6.80°) \text{ A}$$

Although ω should be in rad/sec to calculate the impedances for the inductor and capacitor, the phase angle can be in radians or, as in this example, in degrees, for phasor calculations of $\underline{I}$ or $\underline{V}$.

Example 3.2: Phasor calculation using superposition

Find $i(t)$ in Fig. 3.10(i) using the superposition theorem.

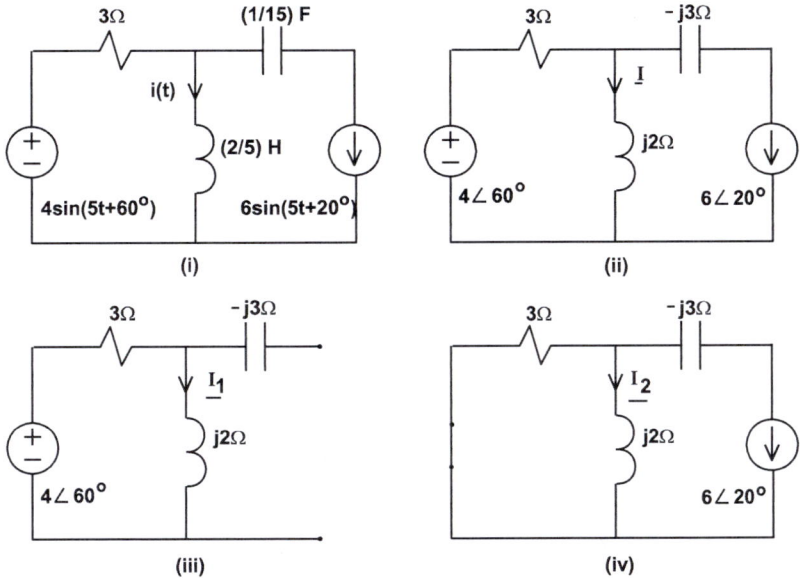

Fig. 3.10 (i)Time and (ii) phasor circuit for Example 3.2.

With the voltage source acting alone, Fig. 3.10(iii) gives

$$I_1 = \frac{4\angle 60°}{(3+j2)} = \frac{4\angle 60°}{3.61\angle 33.69°} = 1.11\angle 26.31° \text{ A}$$

With the current source acting alone, Fig. 3.10(iv) gives

$$I_2 = -\left\{\frac{1/(j2)}{[(1/j2)+(1/3)]}\right\}6\angle 20° = -\left\{\frac{3}{(3+j2)}\right\}6\angle 20°$$

$$I_2 = -\frac{18\angle 20°}{3.61\angle 33.69} = -4.99\angle -13.69° \text{ A}$$

$$I = I_1 + I_2$$

$$i(t) = \{1.11\sin(5t+26.31°) - 4.99\sin(5t-13.69°)\} \text{ A}$$

Example 3.3: Phasor calculation using Thévenin equivalent

Find $\underline{I}$ and $\underline{V}_{ab}$ in Fig. 3.11(i) using Thévenin's theorem.

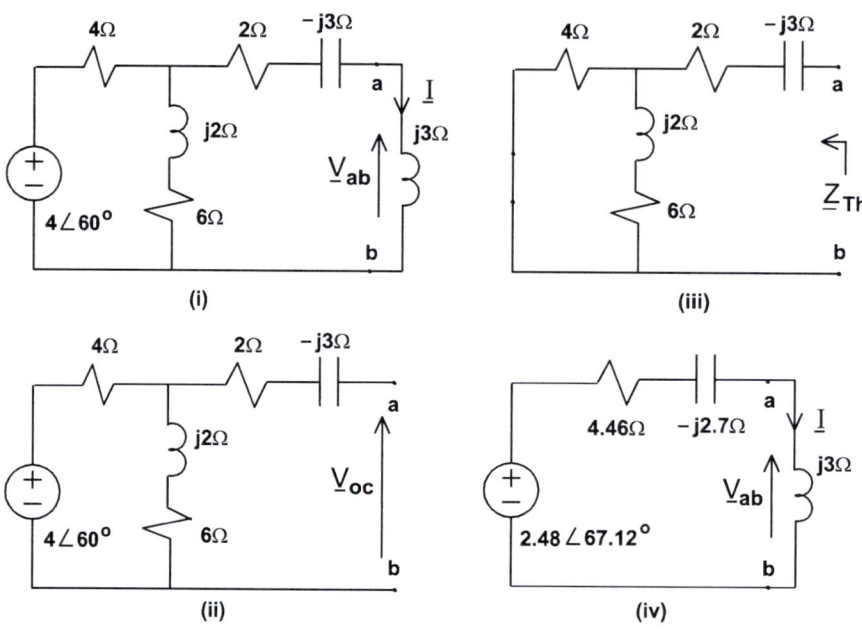

Fig. 3.11. Circuits for Example 3.3.

Removing the j3Ω inductor in Fig. 3.11(i), to find the Thévenin voltage V_{Th} of the circuit to the left of terminals a and b, and using voltage division in Fig. 3.11(ii) gives

$$\underline{V}_{Th} = \underline{V}_{oc} = \left(\frac{6 + j2}{4 + 6 + j2} \right) 4 \angle 60^\circ = \frac{(6.32 \angle 18.43^\circ)(4 \angle 60^\circ)}{10.20 \angle 11.31^\circ}$$

$$\underline{V}_{Th} = (2.48 \angle 67.12^\circ) \text{ V}$$

Looking into terminals a and b with the voltage source deactivated in Fig. 3.11(iii) gives

$$\underline{Z}_{Th} = (2 - j3) + \frac{1}{[(1/4) + 1/(6 + j2)]} = (2 - j3) + \frac{(12 + j4)}{(5 + j)}$$

$$\underline{Z}_{Th} = (2 - j3) + \frac{12.65\angle 18.43°}{5.10\angle 11.31°} = (2 - j3) + 2.48\angle 7.12°$$

$$\underline{Z}_{Th} = (2 - j3) + 2.46 + j0.31 = 4.46 - j2.7 = (5.21\angle -31.19°) \ \Omega$$

Connecting the j3Ω inductor to the Thévenin equivalent circuit results in the circuit of Fig. 3.11(iv) which gives

$$\underline{I} = \frac{2.48\angle 67.12°}{(4.46 - j2.7 + j3)} = \frac{2.48\angle 67.12°}{(4.46 + j0.3)} = \frac{2.48\angle 67.12°}{4.47\angle 3.85°}$$

$$\underline{I} = 0.55\angle 63.27° \ A$$

Ohm's law in Fig. 3.11(iv) gives

$$\underline{V}_{ab} = (0.55\angle 63.27°)(j3) = (0.55\angle 63.27°)(3\angle 90°)$$

$$\underline{V}_{ab} = (1.65\angle 153.27°) \ V$$

Example 3.4: Phasor calculation using single node pair method and current division

Find a) V_{ab} using the single-node-pair method and b) I using current division, in the circuit of Fig. 3.12.

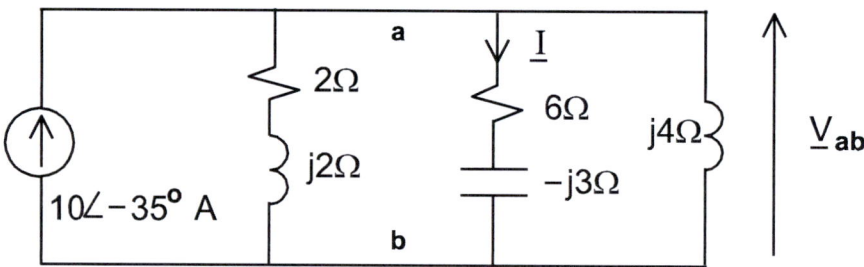

Fig. 3.12. Circuit for Example 3.4.

a) V_{ab}

Ohm's law and KCL give

$$10\angle-35° = \frac{V_{ab}}{(2+j2)} + \frac{V_{ab}}{(6-j3)} + \frac{V_{ab}}{(j4)}$$

$$V_{ab} = \frac{10\angle-35°}{\{[1/(2+j2)]+[1/(6-j3)]+[1/j4]\}}$$

$$V_{ab} = \frac{10\angle-35°}{0.5786\angle-48.504°} = (17.28\angle13.50°) \text{ V}$$

b) I

$$I = \left\{\frac{1/(6-j3)}{[1/(2+j2)]+[1/(6-j3)]+[1/(j4)]}\right\}10\angle-35°$$

$$I = (0.26\angle75.07)(10\angle-35°) = (2.58\angle40.07°) \text{ A}$$

110

Example 3.5: Circuit with sine and cosine sources

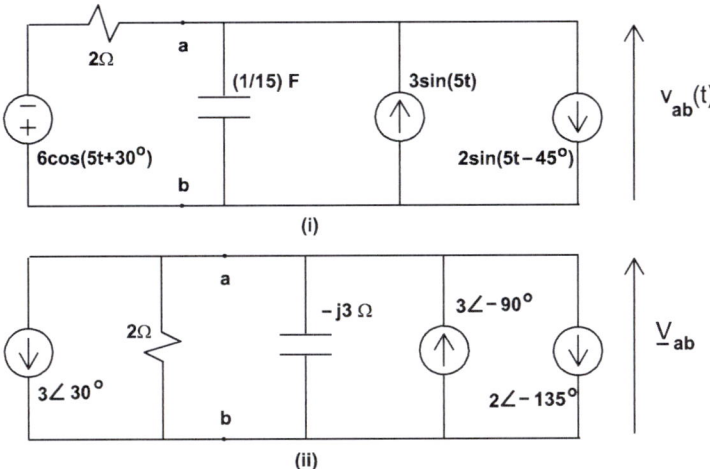

(i)

(ii)

Fig.3.13. Find $v_{ab}(t)$ in (i).Example 3.5.

   The voltage source is a cosine wave while the current sources are sine waves. Sources must either be all sine or all cosine functions. Choosing to use cosine functions, the current sources are converted to cosine functions, according to Equ. 3.10, so that

$$3\sin(5t) = 3\cos(5t - 90°) = (3\angle -90°) \text{ A}$$
$$2\sin(5t - 45°) = 2\cos(5t - 45° - 90°) = (2\angle -135°) \text{ A}.$$

   Then, converting the circuit to the left of terminals a and b in Fig. 3.13(i), to its Norton equivalent in Fig. 3.13(ii), the single-node-pair method for the phasor circuit of Fig. 3.13(ii) gives

$$(3\angle -90°) - (2\angle -135°) - (3\angle 30°) = (V_{ab}/2) + \{V_{ab}/(-j3)\}$$
$$V_{ab} = \frac{3.305\angle -110.99°}{0.601\angle 33.69°} = (5.50\angle -144.68°) \text{ V}$$
$$v_{ab}(t) = 5.5\cos(5t - 144.68°) \text{ V}.$$

111

Example 3.6: Circuit with sources of different frequencies

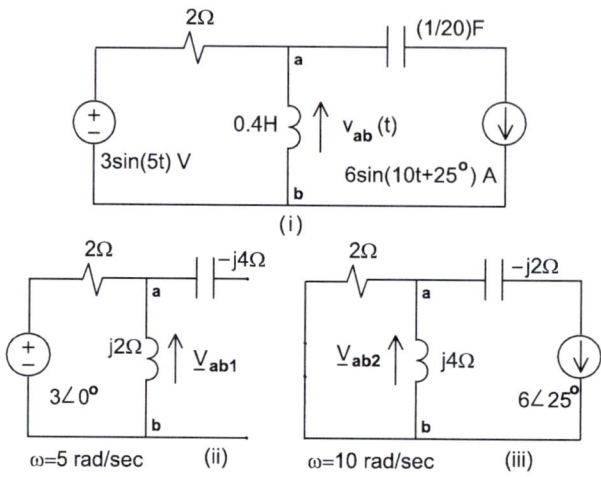

Fig. 3.14. Circuits for Example 3.6. Find $v_{ab}(t)$ in (i).

The angular frequencies of the sources in this circuit are different. Therefore, $j\omega L$ and $1/j\omega C$ have different values at the two frequencies. Superposition is used to find the component of voltage due to a source acting alone, in a circuit with impedances calculated using the frequency of that source. Fig. 3.14(ii) shows the circuit for the voltage source active and the current source deactivated. The current source is active in Fig. 3.14(iii). If $\underline{V}_{ab}$ is the phasor representation of $v_{ab}(t)$, and $\underline{V}_{ab1}$ and $\underline{V}_{ab2}$ are the components of $\underline{V}_{ab}$ due to the voltage source and the current source, respectively, then from Figs. 3.14(ii) and (iii)

$$\underline{V}_{ab1} = \left\{ \frac{j2}{j2+2} \right\} 3\angle 0° = (2.12\angle 45°) \text{ V}$$

$$\underline{V}_{ab2} = -j4 \left\{ \frac{1/(j4)}{[1/(j4)]+1/2} \right\} 6\angle 25° = -(10.73\angle 51.57°) \text{ V}$$

$$v_{ab}(t) = \{2.12\sin(5t + 45°) - 10.73\sin(10t + 51.57°)\} \text{ V}$$

Example 3.7: Circuit with ac and dc sources

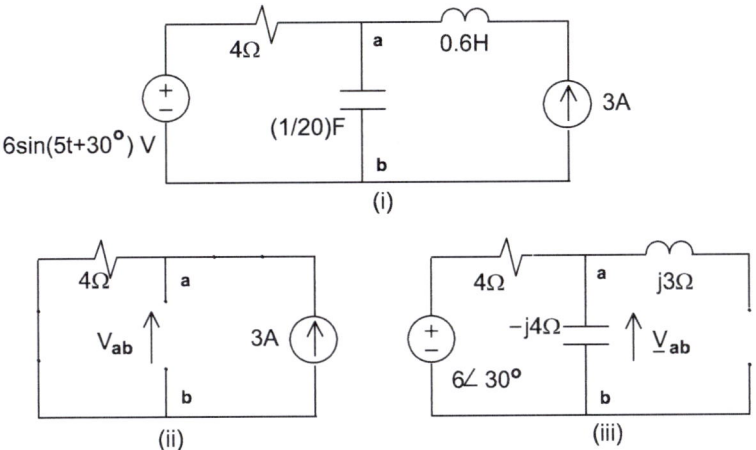

Fig. 3.15 Circuits for Example 3.7. Find $v_{ab}(t)$ in (i).

The circuit of Fig. 3.15(i) contains a dc current source and an ac voltage source. The components of $v_{ab}(t)$ due to each source will be determined using superposition. Fig. 3.15(ii) shows the circuit under dc conditions with the 3A source active, the inductor and capacitor replaced by a short circuit and an open circuit, respectively and with the ac source deactivated. Then the dc component of $v_{ab}(t)$ is given by Ohm's law as $V_{ab} = 3 \times 4 = 12V$.

Fig. 3.15(iii) shows the circuit with the ac voltage source active, the dc source deactivated and with impedances calculated for $\omega = 5$ radians/sec. Then voltage division gives the ac component of $v_{ab}(t)$ as

$$\underline{V}_{ab} = \{(-j4)/(4-j4)\} \times (6\angle 30°) = (4.24\angle -15°) \text{ V}$$

Then

$$v_{ab}(t) = V_{ab} + \underline{V}_{ab} = \{12 + 4.24\sin(5t - 15°)\} \text{ V}$$

113

Example 3.8: Node voltage method in ac circuit

Find $\underline{V}_{ac}, \underline{V}_{bc}$ and $\underline{I}$ using the node voltage method.

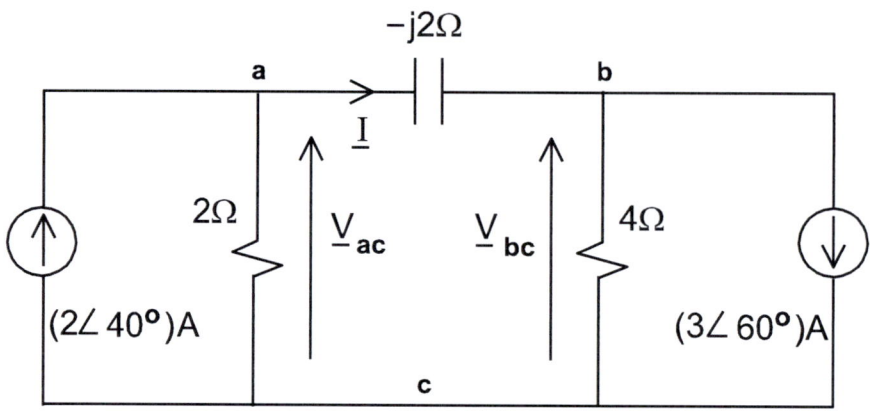

Fig. 3.16. Circuit for Example 3.8.

KCL at nodes a and b and KVL in loop abc give

$$2\angle 40° = \left(\frac{\underline{V}_{ac}}{2}\right) + \underline{I}$$

$$\underline{I} = \left(\frac{\underline{V}_{bc}}{4}\right) + 3\angle 60°$$

$$\underline{V}_{ac} - \underline{V}_{bc} = \underline{I}(-j2)$$

Phasor algebra then gives

$$\underline{V}_{ac} = (2.76\angle -53.58°) \text{ V}$$
$$\underline{V}_{bc} = (3.25\angle -165.74°) \text{ V}$$
$$\underline{I} = 2.50\angle 73.47° \text{ A}$$

114

Example 3.9: Node voltage method in ac circuit

Find $\underline{V}_{ac}$ and $\underline{V}_{bc}$ using the node voltage method.

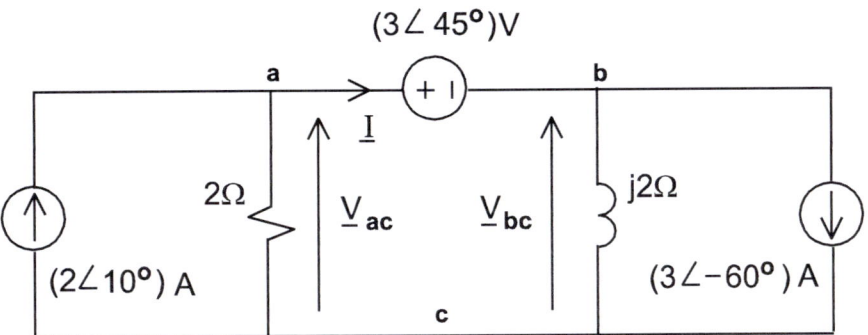

Fig. 3.17. Circuit for Example 3.9.

KCL at nodes a and b and KVL in loop abc give

$$2\angle 10 = \left(\frac{\underline{V}_{ac}}{2}\right) + \underline{I}$$

$$\underline{I} = \left(\frac{\underline{V}_{bc}}{j2}\right) + 3\angle -60$$

$$\underline{V}_{ac} = \underline{V}_{bc} + 3\angle 45°$$

Phasor algebra then gives

$$\underline{V}_{ac} = (3.43\angle 95.93°) \text{ V}$$
$$\underline{V}_{bc} = (2.79\angle 152.41°) \text{ V}$$

Chapter 4
Power in ac circuits

4.1 Introduction

Electrical energy supplied to a resistor by an ac source is converted to heat and the power supplied to the resistor is referred to as average or true power. Average power is measured in watts, W. Ac Power supplied to a reactive element, such as an inductor or capacitor, is stored in the element over part of a cycle and then returned to the supply. This power is not absorbed as heat; it is referred to as reactive power. Reactive power is measured in volts-amperes reactive, VAR. The concept of real and reactive power is discussed in Chapter 4 through the use of instantaneous power and then by using the formula for phasor power. The formula for phasor power is used to calculate average, reactive, and apparent power. In an ac circuit with a load that contains resistive and reactive element, some power is absorbed as heat in the load and some stored and then returned to the supply. The ability of a load to absorb power is measured by its power factor. The power factor of a load can be calculated from the formula for phasor power or from the power triangle

4.2 Power absorbed by resistor; average power

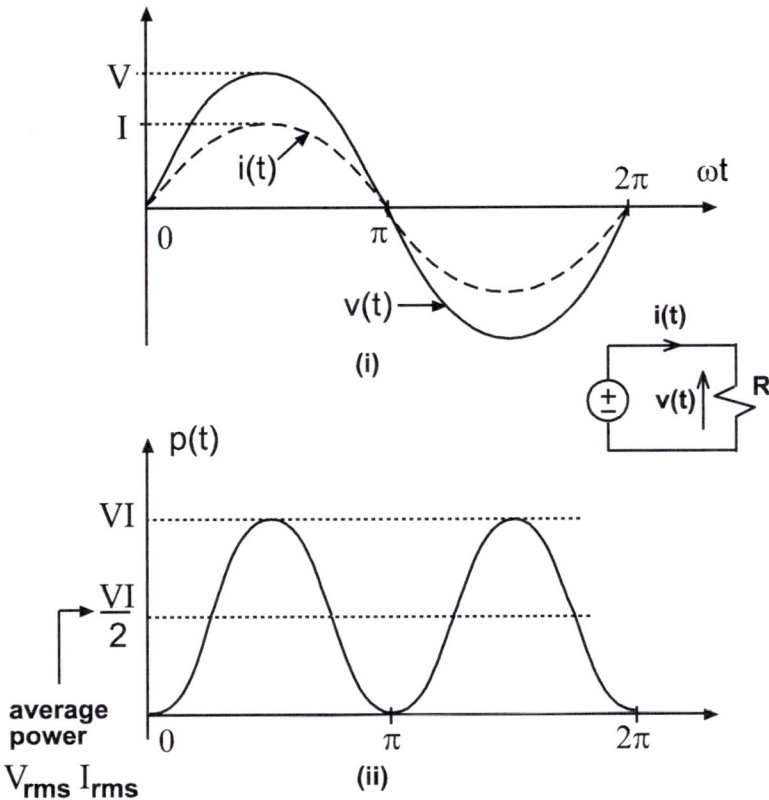

Fig. 4.1 (i)Voltage $v(t)$, current $i(t)$ and (ii) instantaneous power
$p(t)$ associated with a resistor.

Let $v(t)$ be the voltage applied across a resistor, and $i(t)$ the resulting current flowing through it. Then $v(t)$ and $i(t)$ can be expressed as

$$v(t) = V \sin(\omega t) \tag{4.1}$$
$$i(t) = I \sin(\omega t) \tag{4.2}$$

117

where V and I are the voltage and current amplitudes. Equs. 4.1 and 4.2 are plotted in Fig. 4.1(i) and reflect the fact that the voltage and current are in phase. Then, at any time t, the instantaneous power to the resistor, in Watts (abbreviated W), is

$$p(t) = v(t)i(t) = VI\sin^2(\omega t) = \frac{VI}{2}[1 - \cos(2\omega t)] \tag{4.3}$$

The plot of $p(t)$, in Fig. 4.1(ii), shows that the power is always positive. This means that there is a unidirectional flow of power; from the source to the resistor. Energy flowing into the resistor is converted to heat. The current flows into the positive terminal of the resistor at all times and the resistor acts as an absorber of energy. The power varies from zero to VI. The average of the periodic waveform representing the instantaneous power plotted in Fig. 4.1 is the area under the curve over a period, divided by the period, and equals $VI/2$. Since $V_{rms} = V/\sqrt{2}$ and $I_{rms} = I/\sqrt{2}$ then the average power supplied to the resistor is

$$P_{av} = \frac{VI}{2} = V_{rms}I_{rms} \tag{4.4}$$

Since from Ohm's law $V_{rms}=I_{rms}R$, then the average power flowing to the resistor of R ohms can also be written as

$$P_{av} = I_{rms}^2 R \tag{4.5}$$
$$P_{av} = V_{rms}^2 / R \tag{4.6}$$

Average power is also known as true, real or active power and is measured in watts. Equs. 4.4 to 4.6 are the same as for dc power absorbed by a resistor except that now dc voltage and current have been replaced by rms voltage and current.

4.3 Power in inductor; reactive power

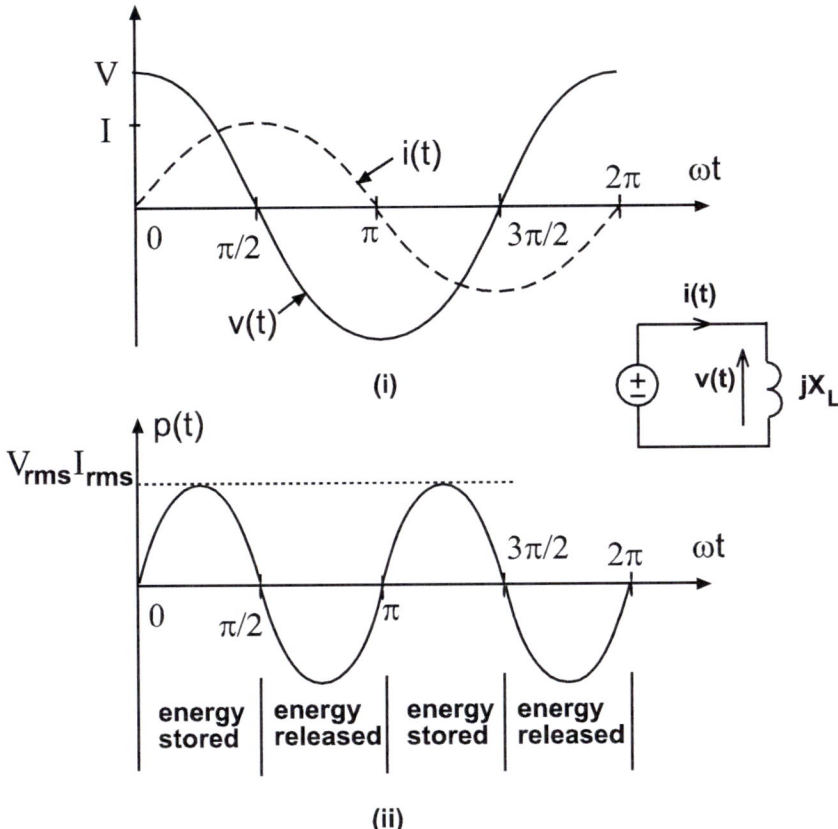

Fig. 4.2 (i) voltage $v(t)$ across and current $i(t)$ through an inductor and (ii) instantaneous power $p(t)$ associated with the inductor.

Consider a sinusoidal voltage source applied across an inductor so that a voltage $v(t)$ appears across the inductor and a current $i(t)$ flows through the inductor. Since the voltage leads the current by $\pi/2$ for the inductor

$$i(t) = I\sin(\omega t) \tag{4.7}$$

$$v(t) = V\sin(\omega t + 90°) \tag{4.8}$$

The instantaneous current is then

$$p(t) = v(t)i(t) = VI \sin(\omega t)\sin(\omega t + 90°) = VI \sin(\omega t)\cos(\omega t) \quad (4.9)$$

$$p(t) = \left(\frac{VI}{2}\right)\sin(2\omega t) \quad (4.10)$$

$$p(t) = V_{rms}I_{rms}\sin(2\omega t) \quad (4.11)$$

Equ. 4.11 is plotted in Fig. 4.2(ii). The plot shows that the power is positive for the first quarter of the cycle and negative over the next quarter cycle. During the first quarter cycle the inductor acts as an absorber of power; the voltage and current directions are as indicated in the circuit in Fig. 4.2 with current flowing into the positive terminal of the inductor. In the next quarter cycle the inductor acts as a generator; the current direction is as indicated in the circuit but the voltage arrow is now pointing down and current flows out of the positive terminal of the inductor. The inductor alternates between storing electrical energy and then releasing this energy back to the voltage supply. The power flowing into and out of the inductor is called reactive power. The average power over a complete cycle is zero and a pure inductor dissipated no energy.

$V_{rms}I_{rms}$ in Equ. 4.11 is referred to as reactive power and given the symbol Q_L. Reactive power is measured in units of volt-amperes reactive, VAR. The impedance, or complex resistance, of an inductor is

$$\underline{Z}_L = j\omega L = jX_L \quad (4.12)$$

where the magnitude $X_L = \omega L$, of the impedance, is the reactance of the inductor. The reactance X_L is analogous to resistance and Ohm's law gives $V_{rms}=I_{rms}X_L$. Reactive power can therefore be written as

$$Q_L = V_{rms}I_{rms} = I_{rms}^2 X_L = V_{rms}^2 / X_L \quad (4.13)$$

120

4.4 Power in capacitor

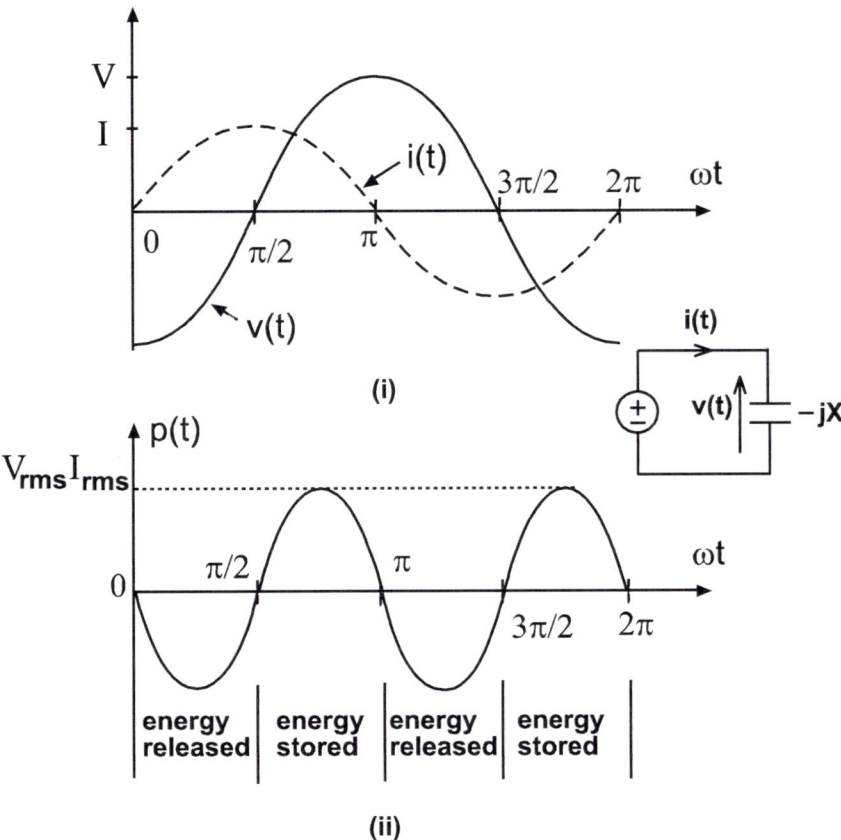

(i)

(ii)

Fig. 4.3 (i) voltage $v(t)$ across and current $i(t)$ through a capacitor and (ii) instantaneous power $p(t)$ associated with the capacitor.

The voltage $v(t)$ across the capacitor lags the capacitor current $i(t)$ by 90°. These waveforms are sketched in Fig. 4.3(i) and can be written as

$$i(t) = I\sin(\omega t) \tag{4.14}$$
$$v(t) = V\sin(\omega t - 90°) \tag{4.15}$$

The instantaneous power is then

$$p(t) = v(t)i(t) = VI \sin(\omega t)\sin(\omega t - 90°) \tag{4.16}$$

$$p(t) = -VI \sin(\omega t)\cos(\omega t) \tag{4.17}$$

$$p(t) = -\frac{VI}{2}\sin(2\omega t) \tag{4.18}$$

$$p(t) = -V_{rms}I_{rms}\sin(2\omega t) \tag{4.19}$$

The power is plotted in Fig. 4.3(ii). The average power is zero and no energy is dissipated. Reactive power flows out of the capacitor for the first quarter of the cycle, when the capacitor acts as a generator of energy. The capacitor is an energy absorber over the second quarter cycle. The reactive power flowing in the circuit is defined as $Q_c = -V_{rms}I_{rms}$ and has the units of VAR. Reactive power is taken to be negative for a capacitor and positive for an inductor.

The impedance, or complex resistance, of a capacitor is

$$\underline{Z}_C = 1/(j\omega C) = -jX_C \tag{4.20}$$

where the magnitude $X_C = 1/(\omega C)$, of the impedance, is the reactance of the capacitor. The reactance X_C is analogous to resistance and Ohm's law gives $V_{rms}=I_{rms}X_C$. Reactive power for the capacitor can therefore be written as

$$Q_C = -V_{rms}I_{rms} = -I_{rms}^2 X_C = -V_{rms}^2/X_C \tag{4.21}$$

4.5 Phasor power

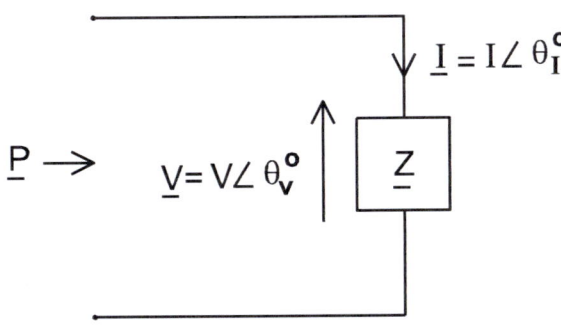

Fig. 4.4 Impedance $\underline{Z}$ represents a general load, which can contain resistive as well as inductive and capacitive elements, that has a voltage $\underline{V}$ across it and a current $\underline{I}$ passing through it. The phasor power associated with the load is $\underline{P}$.

A load, in general, can have a resistive and a reactive component, and the energy source will then supply both active and reactive power to it. The active power will result in heating, or in doing work, and the reactive power will be stored over part of a cycle and then returned to the source. The active and reactive power flowing to the load can be calculated using the equation for phasor power. If the voltage across the load of impedance $\underline{Z}$ is $\underline{V}$ and the current through the load is $\underline{I}$, then the phasor power for the load is

$$\underline{P} = \frac{\underline{V}\,\underline{I}^*}{2} = \underline{V}_{rms}\,\underline{I}^*_{rms} \qquad (4.22)$$

$\underline{I}^*$ is the complex conjugate of $\underline{I}$ so that if $\underline{I} = I\angle\theta_I$, then $\underline{I}^* = I\angle -\theta_I$, or if $\underline{I} = a + jb$, then $\underline{I}^* = a - jb$. Also, if $\underline{V} = V\angle\theta_V$, then $\underline{V}_{rms} = (V/\sqrt{2})\angle\theta_V$ and $\underline{I}^*_{rms} = (I/\sqrt{2})\angle -\theta_I$, where V and I are amplitudes. Equ. 4.22 then gives

$$\underline{P} = V_{rms} \angle \theta_V I_{rms} \angle -\theta_I \qquad (4.23)$$

$$\underline{P} = V_{rms} I_{rms} \angle (\theta_V - \theta_I) \qquad (4.24)$$

$$\underline{P} = V_{rms} I_{rms} [\cos(\theta_V - \theta_I) + j\sin(\theta_V - \theta_I)] \qquad (4.25)$$

$$\underline{P} = P_{av} + jQ \qquad (4.26)$$

The real part, P_{av}, of the phasor power gives the average power in watts and the complex part, Q, is the reactive power measured in volt-ampere reactive (VAR) where

$$P_{av} = V_{rms} I_{rms} \cos(\theta_V - \theta_I) \qquad (4.27)$$

$$Q = V_{rms} I_{rms} \sin(\theta_V - \theta_I) \qquad (4.28)$$

The product $V_{rms}I_{rms}$, in the equation for phasor power, is the apparent power, measured in volt-amperes (VA) and is given the symbol S. An impedance with a voltage V_{rms} across it and a current I_{rms} through it appears to have power of $V_{rms}I_{rms}$ flowing to it, but this is neither average nor reactive power. Since, by analogy to dc power, $V_{rms}I_{rms}$ appears to represent power, it is termed apparent power. The concept of phasor, active, reactive and apparent power applies to the energy source, an impedance representing combinations of resistive and reactive elements and to individual elements in an electrical circuit.

In any ac circuit phasor power is conserved. This means that for each type of power, whether phasor, average or reactive, the power supplied by the energy source equals the sum of the power delivered to the individual elements in the circuit. Conservation does not apply to apparent power.

4.6 Power equation applied to circuit elements

If the load is a pure resistor of resistance R ohms, then the voltage and current are in phase and $(\theta_V - \theta_I) = 0$. Equ. 4.25 gives

$$\underline{P} = V_{rms} I_{rms} [\cos(0) + j\sin(0)] = V_{rms} I_{rms} = P_{av} + jQ$$
$$P_{av} = V_{rms} I_{rms} = V_{rms}^2 / R = I^2 R \tag{4.28}$$
$$Q = 0$$

This gives the power dissipated as heat by the resistor and shows that the reactive power for a pure resistor is zero.

For a pure inductor as the load, the voltage leads the current by $90°$ and so $(\theta_V - \theta_I) = 90°$. Equ. 4.25 then gives

$$\underline{P} = V_{rms} I_{rms} [\cos(90°) + j\sin(90°)] = jV_{rms} I_{rms} = P_{av} + jQ$$
$$P_{av} = 0$$
$$Q_L = V_{rms} I_{rms} \tag{4.29}$$

This shows that a pure inductor absorbs no real or average power but stores and then releases reactive power.

If the load is a pure capacitor the voltage lags the current by $90°$ and $(\theta_V - \theta_I) = -90°$. The phasor power equation then gives

$$\underline{P} = V_{rms} I_{rms} [\cos(-90°) + j\sin(-90°)] = -jV_{rms} I_{rms} = P_{av} + jQ$$
$$P_{av} = 0$$
$$Q_C = -V_{rms} I_{rms} \tag{4.30}$$

This means that a pure capacitor dissipates no real or average power but stores and releases reactive power. The phasor power equation gives a negative reactive power for the capacitor and a positive one for the inductor.

4.7 Power triangle and power factor

Displaying the real and imaginary parts of phasor power on the complex plane results in the power triangle. The impedance $\underline{Z}$ in Fig. 4.5 has a voltage $\underline{V}$ appearing across it and a

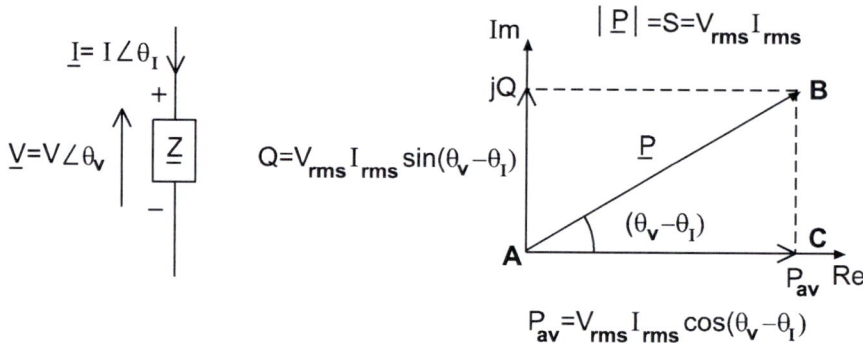

Fig. 4.5 Power triangle showing the real and imaginary components of phasor power.

current $\underline{I}$ passing through it, where V and I are amplitudes. The phasor power for the impedance is

$$\underline{P} = \underline{V}_{rms}\underline{I}_{rms}\cos(\theta_V - \theta_I) + j\underline{V}_{rms}\underline{I}_{rms}\sin(\theta_V - \theta_I) \qquad (4.31)$$

Drawing the real and imaginary components of $\underline{P}$ gives triangle ABC of Fig. 4.5. Side AC is the average power P_{av} drawn along the real axis. Side BC is the reactive power Q drawn parallel to the imaginary axis. The length of the hypotenuse gives the magnitude of $\underline{P}$ or $S=V_{rms}I_{rms}$. ABC is called the power triangle. The angle between $\underline{P}$ and P_{av} is $(\theta_V - \theta_I)$ and the cosine of this angle is called the power factor, pf, of the impedance so that

$$pf = \cos(\theta_V - \theta_I) \qquad (4.32)$$

126

$$pf = \frac{P_{av}}{|P|} = \frac{P_{av}}{S} = \frac{P_{av}}{V_{rms}I_{rms}} \tag{4.33}$$

The power factor of the impedance has a value between 0 and 1 and is the measure of the ability of the impedance to absorb average power. Consider a source connected to impedance $\underline{Z}$. If $\underline{Z}$ is purely resistive $(\theta_V - \theta_I) = 0$, pf=1 and maximum average power is delivered to the impedance. The impedance is then referred to as having unity power factor. If $\underline{Z}$ is purely reactive $(\theta_V - \theta_I) = \pm 90°$, pf=0 and the source delivers no average power to $\underline{Z}$. If $\underline{Z}$ has a resistive and a reactive component, 0< pf<1 and a mixture of active and reactive power flows to $\underline{Z}$, but the active power is less than when $\underline{Z}$ is purely resistive.

Equ. 4.33 shows that the power factor of $\underline{Z}$ is the ratio of average power to apparent power. For unity power factor for instance, active power equals apparent power.

If $\underline{Z}$ represents an inductive load the reactive power Q will be positive and the power triangle will be in the first quadrant, as in Fig. 4.5. Also, for an inductive $\underline{Z}$ the current through $\underline{Z}$ lags the voltage across it and $\underline{Z}$ has a lagging power factor. If $\underline{Z}$ represents a capacitive load, Q will be negative and the power triangle will be in the fourth quadrant. For a capacitive $\underline{Z}$ the current through $\underline{Z}$ leads the voltage across it and $\underline{Z}$ is referred to as having a leading power factor. So this characterization of pf depends on whether the current for $\underline{Z}$ lags or leads the voltage.

The power triangle can be constructed by adding P_{av} and Q for individual elements, irrespective of how they are connected, whether in series or parallel, to find the total P_{av} and total Q that are entered in the complex plane to get the power triangle. In the adding process Q is positive for an inductive element and negative for a capacitive one.

4.8 Maximum power transfer

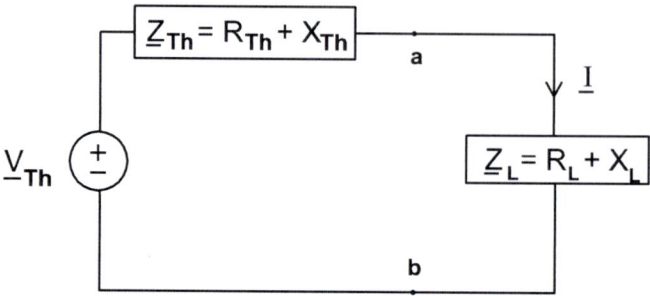

Fig. 4.6 Ac power supply to left of terminals a and b, represented by Thévenin equivalent circuit, delivering power to load of impedance $\underline{Z}_L$.

An ac power supply, represented by its Thévenin voltage and impedance shown to the left of terminals a and b in Fig. 4.6, supplies power to the load $\underline{Z}_L$. The condition is needed under which maximum average power is delivered to the load impedance. The current $\underline{I}_{rms}$ through $\underline{Z}_L$, and the average power P_{av} delivered to $\underline{Z}_L$, are given by

$$\underline{I}_{rms} = \frac{\underline{V}_{Th,rms}}{(R_{Th} + R_L) + j(X_{Th} + X_L)} \tag{4.34}$$

$$P_{av} = \frac{V_{Th,rms}^2 R_L}{(R_{Th} + R_L)^2 + (X_{Th} + X_L)^2} \tag{4.35}$$

It is assumed that the supply voltage $\underline{V}_{Th}$ and impedance $\underline{Z}_{Th}$ are fixed, and that R_L and X_L can be varied to maximize P_{av}. This can be achieved by minimizing the denominator in Equ. 4.35. X_L can be positive or negative and choosing $X_L = -X_{Th}$ reduces the sum of the reactance terms in the denominator to zero and P_{av} becomes

128

$$P_{av} = \frac{V_{Th,rms}^2 R_L}{(R_{Th} + R_L)^2} \qquad (4.36)$$

This is the same as Equ. 1.121 of Chapter 1, for power transfer to a purely resistive load in a dc circuit. Differentiating P_{av} in Equ. 4.36 with respect to R_L, as in Equs. 1.122 to 1.124 before, and equating dP_{av}/dR_L to zero results in $R_L=R_{Th}$. Therefore, maximum average power is transferred to the load when the load impedance equals the complex conjugate of the Thévenin impedance of the supply

$$\underline{Z}_L = \underline{Z}_{Th}^* \qquad (4.37)$$

$$\underline{Z}_L = R_L + jX_L = R_{Th} - jX_{Th} \qquad (4.38)$$

The load is then matched to the supply and the power is, from Equ. 4.36, given as

$$P_{av,\max} = \frac{V_{Th,rms}^2}{4R_{Th}} \qquad (4.39)$$

4.9 Power calculation with multiple ac sources

Superposition can be used in circuits with multiple power sources varying with the same frequency to find an unknown current or voltage as the sum of individual currents and voltages. The power relationship is not a linear one, however, as power is proportional to the square of current or of voltage. Therefore, in calculating power, the individual currents, for instance, are first found and then added to find the total current, and that total current is used in the $(I_{rms,total}^2 \times R)$ relationship to find the power due to all sources acting together. The total power is not the sum of $I_{rms,i}^2 R$ values, where $I_{rms,i}$ is the current due to the ith source acting alone.

Example 4.1: Ac power, series circuit

Find the average and reactive power of the load. The current source given is in rms.

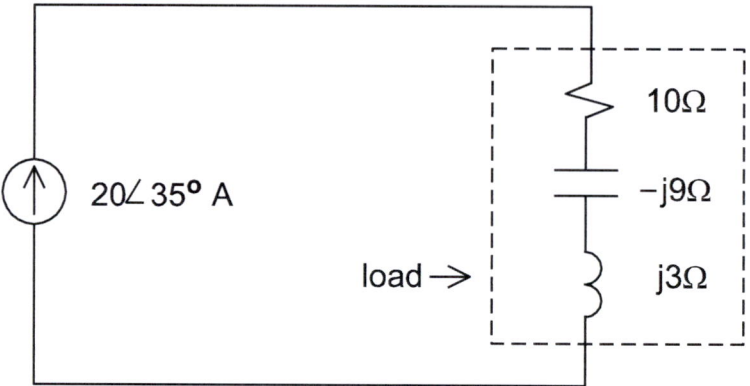

Fig. 4.7 Circuit for Example 4.1.

$$P_{av,10\Omega} = I_{rms}^2 R = 20^2 \times 10 = 4000 \text{ W}$$

$$Q_C = -I_{rms}^2 X_C = -20^2 \times 9 = -3600 \text{ VAR}$$

$$Q_L = I_{rms}^2 X_L = 20^2 \times 3 = 1200 \text{ VAR}$$

Average power $P_{av} = 4000$ W

Total reactive power $Q_T = 1200 - 3600 = -2400$ VAR (capacitive load).

Example 4.2: Ac power, parallel circuit

Find the average and reactive power of the load. The voltage source given is in rms.

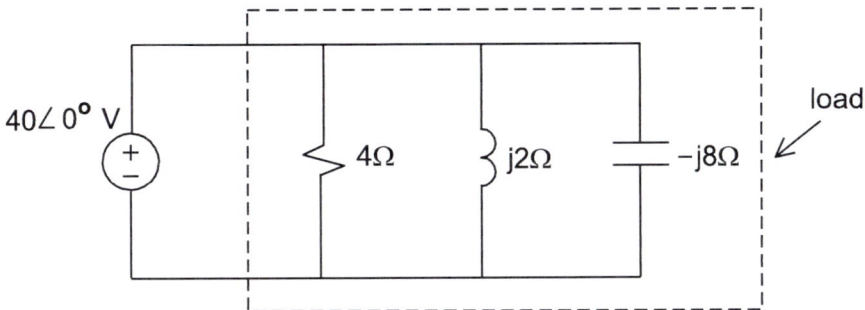

Fig. 4.8 Circuit for Example 4.2.

$$P_{av,4\Omega} = \frac{V_{rms}^2}{R} = \frac{40^2}{4} = 400 \ \text{W}$$

$$Q_L = \frac{V_{rms}^2}{X_L} = \frac{40^2}{2} = 800 \ \text{VAR}$$

$$Q_C = \frac{V_{rms}^2}{X_C} = \frac{40^2}{8} = -200 \ \text{VAR}$$

Average power $P_{av} = 400$ W

Total reactive power $Q_T = 800 - 200 = 600$ VAR (inductive load).

Example 4.3: Ac power, power triangle

The voltage source is in rms. Find the magnitude of the current, I_{rms}, supplied by the source.

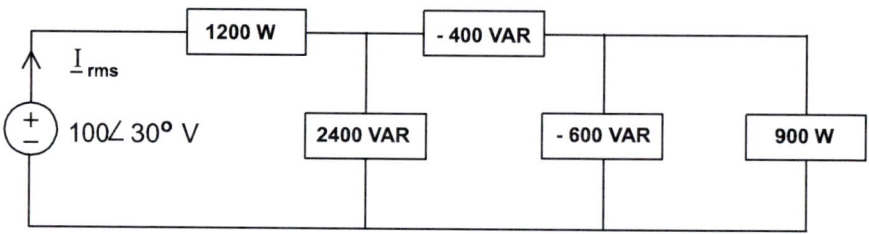

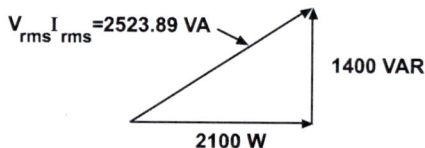

Fig. 4.9 Circuit for Example 4.3.

The total average power and the total reactive power are

$$P_{av,T} = 1200 + 900 = 2100 \text{ W}$$
$$Q_T = 2400 - 400 - 600 = 1400 \text{ VAR (inductive).}$$

The length of the hypotenuse, in the resulting power triangle, is $V_{rms}I_{rms}$.

$$V_{rms}I_{rms} = 2523.89 \text{ VA}$$
$$I_{rms} = \frac{2523.89}{100} = 25.24 \text{ A}$$

Example 4.4: Phasor, average and reactive power

Find the average power P_{av}, and reactive power Q that the voltage source supplies to the load and the power factor pf of the load. The load represents a motor. The source voltage is in rms.

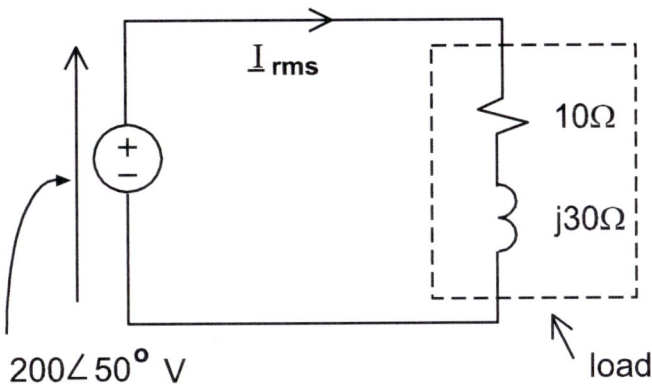

Fig. 4.10 Circuit for Example 4.4.

$$\underline{I}_{rms} = \frac{200\angle 50°}{10 + j30} = 6.325\angle -21.565° \text{ A}$$

$$\underline{P} = \underline{V}_{rms}\underline{I}^{*}_{rms} = (V_{rms}\angle\theta_V)(I_{rms}\angle-\theta_I) = V_{rms}I_{rms}\angle(\theta_V - \theta_I)$$

$$\underline{P} = (200\angle 50°)\times(6.325\angle 21.565°)$$

$$\underline{P} = 1264.91\angle 71.565° = 400 + j1200$$

$$\underline{P} = P_{av} + jQ$$

$$P_{av} = 400 \text{ W}$$

$$Q = 1200 \text{ VAR}$$

$$\text{pf} = \cos(\theta_V - \theta_I) = \cos[50° - (-21.565°)] = \cos(71.565°)$$

pf = 0.32 lagging, since the load current lags the load voltage.

Example 4.5: Phasor, average and reactive power

Find the average and reactive power supplied to the load in the circuit of Fig. 4.11. Then find the power in each element in the load. The voltage source is in rms.

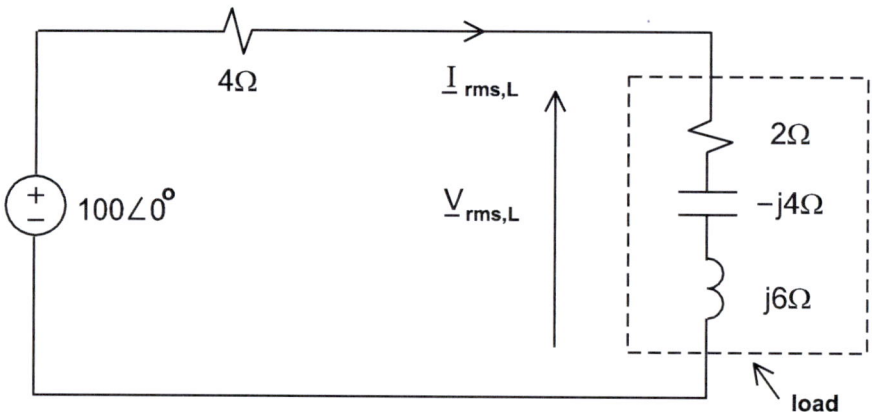

Fig. 4.11 Circuit for Example 4.5.

Voltage division gives the voltage dropped across the load as

$$\underline{V}_{rms,L} = \frac{(2 - j4 + j6)}{(4 + 2 - j4 + j6)} \times (100\angle0°) = 44.721\angle26.565° \text{ V}$$

Using Ohm's law in the loop gives the load current as

$$\underline{I}_{rms,L} = \frac{100\angle0°}{(4 + 2 - j4 + j6)} = 15.811\angle-18.435° \text{ A}$$

Then the phasor power supplied to the load is

$$\underline{P} = \underline{V}_{rms}\underline{I}_{rms}^* = (44.721\angle26.565°)(15.811\angle18.435°)$$
$$\underline{P} = 707.107\angle45° = 500 + j500$$

134

Therefore, the active and reactive powers supplied to the load by the source are

$$P_{av} = 500 \text{ W}$$
$$Q = 500 \text{ VAR}$$

with pf = $\cos(45°) = 0.707$ lagging, as the current lags the voltage.

Considering the power for each element of the load, the average power dissipated in the 2 ohm resistor is

$$P_{av,2\Omega} = I_{rms}^2 R = (15.811)^2 \times 2 = 500 \text{ W}$$

This shows that the average power delivered to the load, by the source, equals the average power dissipated by the resistive element in the load.

The reactive power values for the capacitor and inductor are

$$Q_C = I_{rms}^2 X_C = -(15.811)^2 \times 4 = -1000 \text{ VAR (capacitive)}$$
$$Q_L = I_{rms}^2 X_L = (15.811)^2 \times 6 = 1500 \text{ VAR (inductive)}.$$
$$Q_C + Q_L = 1500 - 1000 = 500 = Q.$$

This shows that the reactive power supplied to the load equals the sum of reactive power values delivered to the reactive elements of the load.

Example 4.6: Phasor, average and reactive power

Find the power delivered by the voltage source to the load. Find the power in each element of the load. The voltage of the source is in rms.

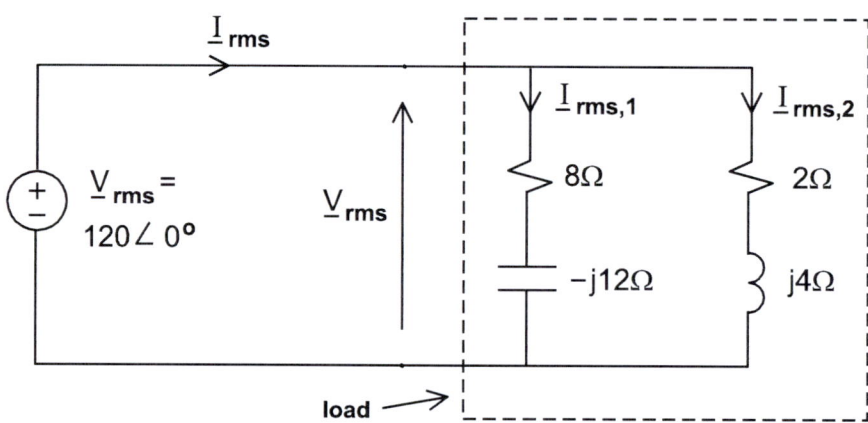

Fig. 4.12 Circuit for Example 4.6.

Ohm's law and KCL give

$$\underline{I}_{rms,1} = \frac{120\angle 0°}{8 - j12} = 8.3205\angle 56.3099° = (4.6154 + j6.9231)\ \text{A}$$

$$\underline{I}_{rms,2} = \frac{120\angle 0°}{2 + j4} = 26.8328\angle -63.4349 = (12 - j24)\ \text{A}$$

$$\underline{I}_{rms} = \underline{I}_{rms,1} + \underline{I}_{rms,2} = (23.8263\angle -45.7848°)\ \text{A}$$

The load has a lagging power factor of

$$\text{pf} = \cos(\theta_V - \theta_I) = \cos(0 - [-45.78°]) = 0.70$$

136

The phasor power supplied by the source is

$$\underline{P} = \underline{V}_{rms}\underline{I}_{rms}^* = P_{av} + jQ$$
$$\underline{P} = (120\angle 0°)(23.8263\angle 45.7848°)$$
$$\underline{P} = 2859.16\angle 45.7848° = 1993.85 + j2049.23$$

Therefore, the average and reactive power values supplied by the source to the load are

$$P_{av} = 1993.85 \ \text{W}$$
$$Q = 2049.23 \ \text{VAR}$$

The average power values delivered to the resistors in the load are

$$P_{av,8\Omega} = I_{rms,1}^2 R = (8.3205^2) \times 8 = 553.846 \ \text{W}$$
$$P_{av,2\Omega} = I_{rms,2}^2 R = (26.8328^2) \times 2 = 1440 \ \text{W}$$
$$P_{av,TOTAL} = 553.846 + 1440 = 1993.84 \ \text{W}$$

The sum of the average power delivered to the resistors in the load equals the average power supplied by the source, showing that average power is conserved.

The reactive power values delivered to the capacitor and inductor of the load are

$$Q_C = -I_{rms,1}^2 X_C = -(8.3205^2) \times 12 = -830.769 \ \text{VAR}$$
$$Q_L = I_{rms,2}^2 X_L = (26.8328^2) \times 4 = 2880 \ \text{VAR}$$
$$Q_{TOTAL} = 2880 - 830.769 = 2049.23 \ \text{VAR}$$

The sum of the reactive power delivered to the capacitor and inductor equals the reactive power supplied by the source, showing that reactive power is conserved.

Example 4.7: Power factor correction

The load, of the circuit in Fig. 4.13, has an rms voltage of 500 V appearing across it, absorbs an average power of 1500W and has a lagging power factor of 0.3.

a) Find the capacitance C of the capacitor which when connected in parallel with the load, corrects the power factor of the combination of the load, shown in the circuit, and the capacitor to unity. Assume that the voltage appearing across the load stays the same after the capacitor is connected across it and that the frequency of the voltage source is 60 Hz.

b) Find the average power dissipated in the transmission line, represented by the resistance of 2 ohms, before and after the capacitor is connected.

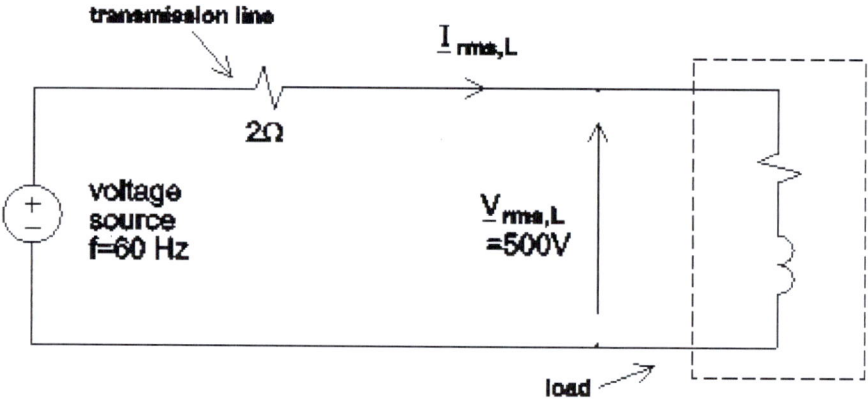

Fig. 4.13 Circuit for Example 4.7.

a) Without the capacitor connected, the magnitude of the load current is

$$I_{rms,L} = \frac{P_{av}}{V_{rms,L} \times \cos(\theta_V - \theta_I)} = \frac{1500}{500 \times 0.3} = 10 \text{ A}$$

$$\theta_V - \theta_I = \cos^{-1} 0.3 = 72.54°$$

Let $\theta_V = 0°$. Then $\theta_I = -72.54°$ and

$$\underline{V}_{rms,L} = (500 \angle 0°) \text{ V}$$
$$\underline{I}_{rms,L} = (10 \angle -72.54°) = (3.000 - j9.539) \text{ A}$$

The reactive power flowing to the load is

$$Q_L = V_{rms,L} I_{rms,L} \sin(\theta_V - \theta_I) = 500 \times 10 \times \sin(72.54°)$$
$$Q_L = 4769.63 \text{ VAR (inductive, since the current lags the voltage)}.$$

The average power dissipated in the transmission line is

$$P_{av,1} = I_{rms,L}^2 \times R = 10^2 \times 2 = 200 \text{ W}$$

b) With the capacitor connected across the load, the reactive power Q_C, flowing into the capacitor, of capacitance C and of reactance X_C, is

$$Q_C = -\frac{V_{rms,L}^2}{X_C} = -\frac{V_{rms,L}^2}{\dfrac{1}{2\pi fC}} = -500^2 \times 2\pi \times 60C = -(9.425 \times 10^7 C)$$

For the load combined with the capacitor to have unity power factor, the reactive power of the capacitor must cancel the inductive reactive power of the load so that

$$Q_C = Q_L$$
$$9.425 \times 10^7 \times C = 4769.63$$
$$C = 50.61\,\mu\text{F}$$

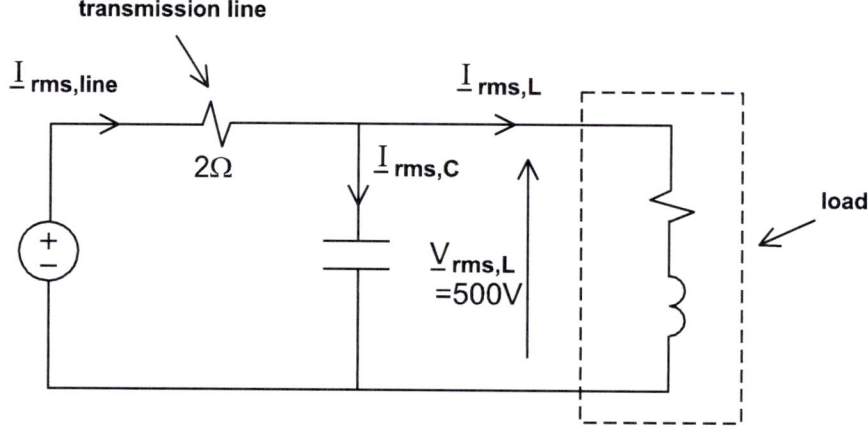

Fig. 4.14. Capacitor of capacitance C connected in parallel with the load to increase power factor to unity. (Example 4.7).

If $\underline{I}_{rms,C}$ is the capacitor current (see Fig. 4.14), then, since the load current is the same as before, the current through the transmission line resistor is given, by KCL, as

$$\underline{I}_{rms,line} = \underline{I}_{rms,C} + \underline{I}_{rms,L}$$
$$\underline{I}_{rms,line} = \frac{\underline{V}_{rms,C}}{\dfrac{1}{j\omega C}} + \underline{I}_{rms,L}$$
$$\underline{I}_{rms,line} = \underline{V}_{rms,C} \times j\omega C + \underline{I}_{rms,L}$$

$$\underline{I}_{rms,line} = 500 \times j \times 2\pi 60 \times 50.61 \times 10^{-6} + 3.000 - j9.539$$

$$\underline{I}_{rms,line} = j9.539 + 3.000 - j9.539$$

$$\underline{I}_{rms,line} = 3 \text{ A}$$

The average power dissipated in the transmission line is now

$$P_{av,2} = I_{rms,line}^{2} \times R_{line} = 3^{2} \times 2 = 18 \text{ W}$$

Since the average power delivered to the capacitor-load combination is the same as without the capacitor, increasing the power factor to unity resulted in a reduction in the current in the transmission line and a reduction in $I^2 R$ power losses. A motor is an inductive load and using a capacitor in parallel with it, to achieve power factor correction, saves energy in electric power systems.

Note that the voltage across, and the current into the capacitor-load combination are in phase, for unity power factor, when the source sees the combination as a purely resistive entity. The reactive power is then supplied to the load by the capacitor over a quarter of a cycle and then returned to the capacitor by the load over the next quarter cycle. The source is then relieved of the task of meeting the reactive power needs of the load. Reactive power is necessary to provide magnetization for motor action.

Chapter 5
The transformer

5.1 Introduction

Two coils of conducting wire, the primary and secondary coils, isolated from one another electrically, but linked together through an alternating magnetic flux that is generated by an alternating current fed into the primary, constitutes a transformer. In the step-up transformer the number of turns, N_2, in the secondary coil is higher than the number of turns, N_1, in the primary and the voltage that appears across the terminals of the secondary is greater than that across the primary. In the step-down transformer $N_2 < N_1$ and the voltage across the secondary is less than that across the primary.

Electrical power is first generated and then transported through transmission lines to locations where it is utilized. Transformers are used at both ends of the transmission line to enable the efficient transport and distribution of ac power.

The transfer of ac power over transmission lines results in loss of average power $I_{rms}^2 R_{line}$, where I_{rms} is the current through the line and R_{line} the line resistance. This energy is wasted as it cannot reach the load to do useful work. The loss in the transmission line can be kept low if the transmission line voltage is high. This follows from the expression for the total average power, $P_{av} = V_{rms} I_{rms} \cos(\theta_V - \theta_I)$, delivered by the energy generator. For a given power factor of $\cos(\theta_V - \theta_I)$ and a given power to be delivered, if V_{rms} is high then I_{rms}, and therefore the line loss, is low. The increase in line voltage is achieved by using a transformer to step up the generated ac voltage to a value, which results in enough reduction in line loss, to enable the economical transfer of power over long distances. At the end of the transmission line transformers are used to, this time, step down the voltage to safer levels that are suitable for local distribution and use.

The ideal transformer will be discussed in this chapter to show how it transforms the voltage and current values at its input to values at its output that depend on the ratio of turns in the primary coil to those in the secondary. Moreover, the transformer transforms the load connected across the secondary to a value, seen at the primary terminals, that depends on the load impedance and the turns ratio. By transforming the impedance of a load, a transformer can be used to achieve maximum power transfer between an ac source and the load. Also, since the transformer only works with ac, it can be used to block the dc component of an input signal and allow the ac component alone to pass through to the output.

5.2 Electric current and magnetic field

An electric current passing through a conducting wire produces a magnetic force field around the conductor. The magnetic field is visualized by drawing closed loops that show the direction of the force that acts on the north pole of a small permanent magnet when the magnet is placed in the field. The lines are called magnetic flux lines. Flux lines are drawn closer together where the magnetic field is strong and further apart where it is weak.

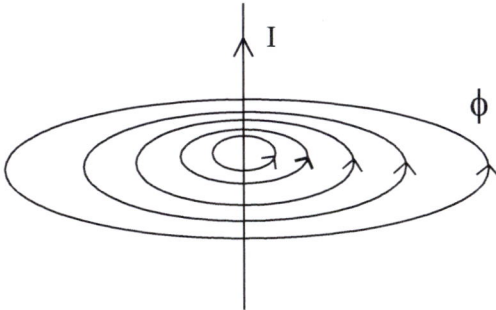

Fig. 5.1 Magnetic flux lines around a current-carrying conductor.

The magnetic flux lines around a current-carrying conductor are concentric rings (see Fig. 5.1). The direction of flux lines, and of the force acting on the north pole of a magnet in the magnetic field, is determined by the right hand rule. The right hand rule states that if the conductor is grasped in the right hand so that the thumb is in the direction of the current, then the fingers curl in the direction of the flux. Flux has the symbol ϕ and is measured in webers (Wb).

The flux density B is the number of flux lines crossing unit cross sectional area, so that if a flux of ϕ webers crosses an area of A meters^2, then the flux density is

$$B = \frac{\phi}{A} \qquad (5.1)$$

B is measured in webers/meter^2 or teslas (T).

For a conductor carrying a constant current of I amperes the flux density at a radial distance r away from the wire is

$$B = \mu_o \mu_r \left(\frac{I}{2\pi r} \right) \qquad (5.2)$$

where B is in a direction tangential to the circle of radius r that surrounds the current, μ_o is the magnetic permeability of free space and equals $4\pi \times 10^{-7}$ henry/meter, and μ_r is the relative permeability of the material surrounding the conductor. For air $\mu_r = 1$, but for a magnetic material such as iron, and for iron alloyed with other ferromagnetic materials, μ_r is over three orders of magnitude higher than μ_r for air. Therefore, the flux density produced by a current-carrying wire will be thousands of times greater in a ferromagnetic material than in air. The property of ferromagnetic materials to concentrate magnetic flux is the reason for their use as the core, of iron-core transformers, on which the primary and secondary coils are wound.

5.3 Magnetomotive force, flux and reluctance in magnetic circuit

When a wire, carrying a current of I amperes, is wrapped around a ferromagnetic material to make a coil of N turns, a magnetic flux of ϕ webers is set up which is concentrated within the ferromagnetic material (see Fig. 5.2). NI is the magnetomotive force (MMF), or the magnetic potential difference, that forces the

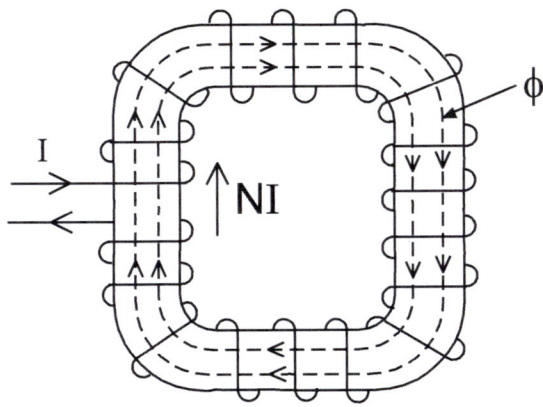

Fig. 5.2 Current-carrying coil wrapped around iron core of magnetic circuit.

flux ϕ around the iron. The MMF is proportional to the current in the wire and the number of turns of the coil

$$MMF = NI \qquad (5.3)$$

The MMF is analogous to the electric potential difference and the flux is analogous to the current in an electric circuit. The flux in Fig. 5.2 flows in a magnetic circuit in which the equivalent of resistance is the reluctance $\Re$. The reluctance is the opposition of the magnetic material to carrying the magnetic flux so that

$$MMF = \Re\phi \qquad (5.4)$$

Equ. 5.4 is analogous to Ohm's law for the magnetic circuit. Reluctance has the units of ampere-turns/weber and is given by

$$\Re = \frac{l}{\mu_o \mu_r A} \qquad (5.5)$$

The reluctance is directly proportional to the magnetic path l, and inversely proportional to the cross sectional area A, of the iron core on which the wire is wound. Since the relative permeability of iron is many times greater than that of air, the reluctance of iron is much smaller than that for air. Therefore, flux lines prefer to concentrate within the iron core rather than the air surrounding it.

5.4 The ideal transformer
5.4.1 Voltage ratio

Fig. 5.3 is a schematic representation of the ideal transformer. It consists of two windings; the primary with N_1 turns and the secondary with N_2 turns. Both coils are wound on a common magnetic circuit called the iron core. For the ideal transformer it is assumed that the reluctance of the core is low enough to allow a negligible MMF to establish the flux. The second assumption is that the flux is confined to the core and passes through, or links, both windings. The resistances of the windings are assumed to be negligible. Lastly, the core losses are assumed to be negligible. Core losses can arise because the changing flux induces voltages in the core and, since the core is a conductor, currents circulate in the iron. This results in the dissipation of energy and in heating. The core of the transformer is made of sheets that are insulated from each other to minimize currents in the core and to keep core losses low.

A sinusoidally varying voltage, $v_1(t)$, is applied across the terminals of the primary winding in the ideal transformer of Fig. 5.3 and results in the primary current $i_1(t)$ which in turn gives the sinusoidally varying magnetic flux $\phi(t) = \phi_{max} \sin(2\pi f t)$ in the iron

146

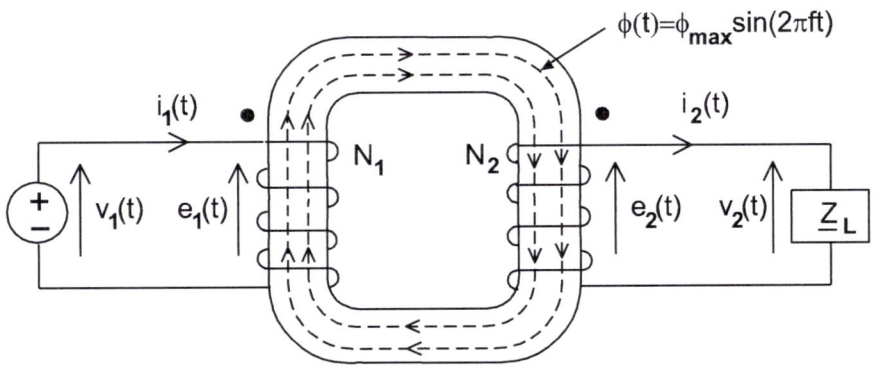

Fig. 5.3 The ideal transformer.

core. The flux links, or passes through, both coils of the transformer and induces a voltage $e_1(t)$ across the primary and a voltage $e_2(t)$ across the secondary coil. The induced voltage is, by Faraday's law, equal to the number of turns in the coil and the rate of change of the flux linking the coil. The voltage induced across the primary coil is therefore given as

$$e_1(t) = N_1 \frac{d\phi(t)}{dt} \tag{5.6}$$

$$e_1(t) = N_1 \frac{d}{dt}\{\phi_{max}\sin(2\pi ft)\} \tag{5.7}$$

$$e_1(t) = 2\pi f N_1 \phi_{max}\cos(2\pi ft) \tag{5.8}$$

$$e_1(t) = E_1 \cos(2\pi ft) \tag{5.9}$$

$$E_1 = 2\pi f N_1 \phi_{max} \tag{5.10}$$

$$E_{rms,1} = \frac{2\pi f N_1 \phi_{max}}{\sqrt{2}} = 4.44 f N_1 \phi_{max} \tag{5.11}$$

where E_1 and $E_{rms,1}$ are, respectively, the amplitude and rms value of $e_1(t)$.

147

Similarly, the voltage induced across the secondary coil is

$$e_2(t) = N_2 \frac{d\phi(t)}{dt} \qquad (5.12)$$

$$e_2(t) = 2\pi f N_2 \phi_{max} \cos(2\pi f t) \qquad (5.13)$$

$$e_2(t) = E_2 \cos(2\pi f t) \qquad (5.14)$$

$$E_2 = 2\pi f N_2 \phi_{max} \qquad (5.15)$$

$$E_{rms,2} = 4.44 f N_2 \phi_{max} \qquad (5.16)$$

where E_2 and $E_{rms,2}$ are, respectively, the amplitude and rms value of $e_2(t)$. Dividing Equs 5.10 and 5.15 gives

$$\frac{E_1}{E_2} = \frac{N_1}{N_2} \qquad (5.17)$$

Equ. 5.17 states that the ratio of the primary voltage to the secondary voltage is equal to the ratio of the number of turns of the primary to the number of turns of the secondary. The primary and secondary voltages can be represented as phasors $\underline{V}_1$ and $\underline{V}_2$. Since, from Fig. 5.3, $v_1(t) = e_1(t)$ and $v_2(t) = e_2(t)$, then

$$\frac{E_1}{E_2} = \frac{E_{rms,1}}{E_{rms,2}} = \frac{\underline{V}_1}{\underline{V}_2} = \frac{V_1}{V_2} = \frac{N_1}{N_2} = a \qquad (5.18)$$

where a is the turns ratio.

The secondary voltage is either in phase, or $180°$ out of phase, with primary voltage, depending on the way the secondary is wound with respect to the primary. The dots indicate the terminals at which the primary and secondary voltage polarities are the same. For the transformer of Fig. 5.3 the dotted terminal is the top terminal for both the primary and secondary coils. This means that both these terminals will be positive at the same time and then both will be negative during the second half of the cycle and $e_1(t)$ and $e_2(t)$ shown in the Figure will be in phase. The voltage arrow across a coil is drawn to point to the terminal marked with the dot. The primary current then flows into the positive terminal of the primary, indicating that the primary coil acts as an absorber of energy, and the secondary current flows out of the positive terminal indicating that the secondary coil acts as a generator. If the secondary is now wound so that its voltage is $180°$ out of phase with that of the primary, then the dot is placed at the top terminal of the primary and the bottom terminal of the secondary.

5.4.2 Current ratio

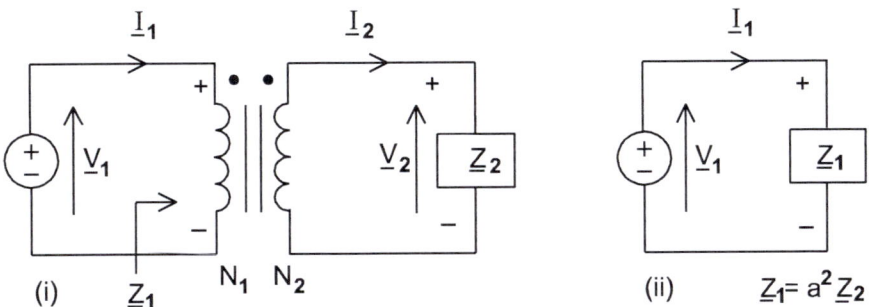

Fig. 5.4 (i) Circuit symbol of the ideal transformer and (ii) $\underline{Z}_1$ the reflected impedance of the load.

Fig. 5.4(i) shows the symbol of the ideal transformer where the parallel lines between the two coils represent the iron core. $\underline{V}_1$ and $\underline{I}_1$ are the primary voltage and current, $\underline{V}_2$ and $\underline{I}_2$ the secondary voltage and current, while $\underline{Z}_2$ is the load impedance connected across the secondary. $\underline{Z}_1$ is the impedance seen when looking into the terminals of the primary coil.

Since there are no losses in the ideal transformer the average power flowing into the primary is equal to the average power flowing out of the secondary. This gives

$$V_1 I_1 = V_2 I_2 \qquad (5.19)$$

where the $\cos(\theta_V - \theta_I)$ terms on either side of Equ. 5.19 are the same and so cancel out. Then, since from Equ. 5.18

$$\frac{V_1}{V_2} = \frac{N_1}{N_2} = a \qquad (5.20)$$

Equ. 5.19 becomes

$$\frac{I_1}{I_2} = \frac{N_2}{N_1} = \frac{1}{a} \qquad (5.21)$$

Therefore, the ratio of the primary current to the secondary current equals the inverse of the turns ratio. In a step-up transformer the voltage is stepped up and the current is stepped down. The currents in Equ. 5.21 can be amplitudes, rms values or phasors and the ratio of currents can then also be given as

$$\frac{\underline{I}_1}{\underline{I}_2} = \frac{1}{a} \qquad (5.22)$$

5.4.3 Reflected impedance

If $\underline{Z}_2$ of Fig. 5.4(i) is the load impedance, that is connected across the terminals of the secondary, then the impedance seen by the source is from Ohm's law

$$\underline{Z}_1 = \frac{V_1}{\underline{I}_1} = \frac{aV_2}{\left(\dfrac{\underline{I}_2}{a}\right)} = a^2\left(\frac{V_2}{\underline{I}_2}\right) \tag{5.23}$$

$$\underline{Z}_1 = a^2\underline{Z}_2 \tag{5.24}$$

The term $a^2\underline{Z}_2$ is called the impedance of the load referred to the primary, or the reflected impedance, and its use transforms the transformer circuit of Fig. 5.4(i) to the single loop circuit of Fig. 5.4(ii).

Example 5.1: Current, voltage and power in transformer

Find $\underline{I}_1, \underline{I}_2, \underline{V}_2$ and the average power supplied by the source in the circuit of Fig.5.5(i). The source voltage is given in rms.

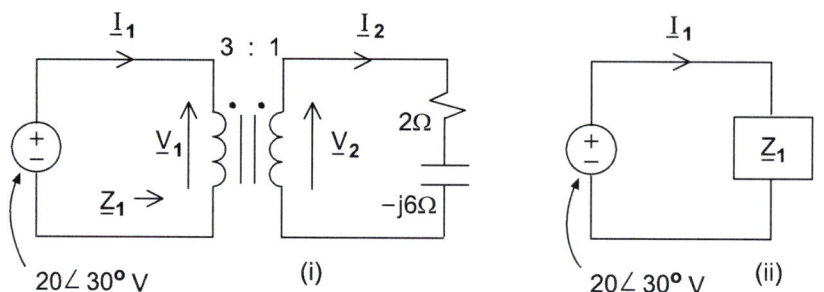

Fig. 5.5 Circuit for Example 5.1.

The impedance of the load referred to the primary is

$$\underline{Z}_1 = a^2 \underline{Z}_2 = (3/1)^2 (2 - j6) = 18 - j54 = (56.92\angle -71.57°)\Omega$$

From Fig. 5.5(ii), Ohm's law and the transformer equations give

$$\underline{I}_1 = (20\angle 30°)/(56.92\angle -71.57°) = (0.35\angle 101.57°) \text{ A}$$

$$\frac{\underline{I}_1}{\underline{I}_2} = \frac{1}{a} = \frac{N_2}{N_1}$$

$$\underline{I}_2 = (3/1)(0.35\angle 101.57) = (1.05\angle 101.57°) \text{ A}$$

$$\underline{P} = \underline{V}_{rms} \underline{I}_{rms}^* = (20\angle 30°)\underline{I}_1^* = (20\angle 30°)(0.35\angle -101.57°)$$

$$\underline{P} = 7.0\angle -71.57° = 2.21 - j6.64 = P_{av} + jQ$$

$$P_{av} = 2.21 \text{ W}$$

$$\frac{\underline{V}_1}{\underline{V}_2} = \frac{N_1}{N_2}$$

$$\underline{V}_2 = (1/3)(20\angle 30°) = (6.67\angle 30°) \text{ V}$$

Example 5.2: Current and voltage in transformer circuit

Find $\underline{V}_{ab}$ and $\underline{V}_1$ in the circuit of Fig. 5.6(i).

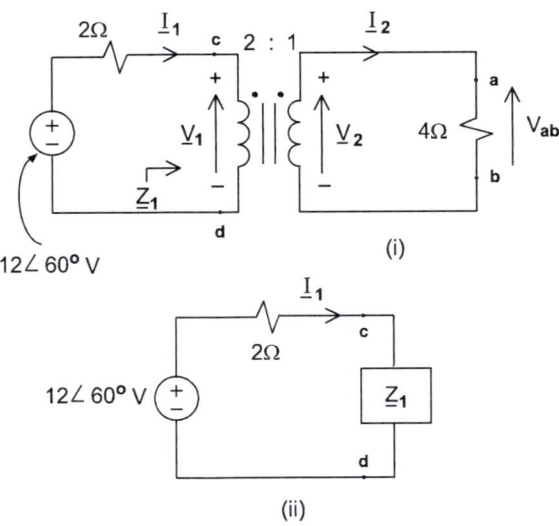

(i)

(ii)

Fig. 5.6 Circuit for Example 5.2.

$\underline{Z}_1 = a^2 4 = 2^2 \times 4 = 16\Omega$ is the reflected impedance. Ohm's law in Fig. 5.6(ii) and the transformer equations give

$$\underline{I}_1 = (12\angle 60°)/18 = (0.667\angle 60°) \text{ A}$$

$$\frac{\underline{I}_2}{\underline{I}_1} = \frac{N_1}{N_2} = \frac{2}{1}$$

$$\underline{I}_2 = 2 \times (0.67\angle 60°) = (1.334\angle 60°) \text{ A}$$

$$\underline{V}_{ab} = 4 \times (1.34\angle 60°) = (5.336\angle 60°) \text{ V}$$

$$\frac{\underline{V}_1}{\underline{V}_2} = \frac{\underline{V}_1}{\underline{V}_{ab}} = \frac{N_1}{N_2}$$

$$\underline{V}_1 = \underline{V}_{ab} \times 2 = (10.67\angle 60°) \text{ V}$$

Example 5.3: Reflected impedances

Find $\underline{V}_{ab}$ in the circuit of Fig. 5.7.

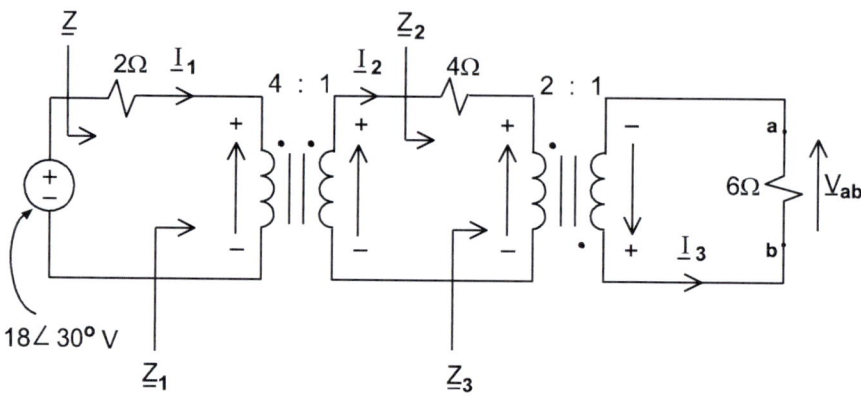

Fig. 5.7 Circuit for Example 5.3.

The load impedance referred to the primary of the 4:1 transformer is $\underline{Z}_1$ where

$$\underline{Z}_3 = (2^2)6 = 24\Omega$$
$$\underline{Z}_2 = 4 + 24 = 28\Omega$$
$$\underline{Z}_1 = (4^2)28 = 448\Omega$$
$$\underline{Z} = 2 + 448 = 450\Omega$$

The source voltage appears across $\underline{Z}$ and Ohm's law and the transformer relationships give

$$\underline{I}_1 = [(18\angle 30°)/(450)] = (0.04\angle 30°) \text{ A}$$
$$\underline{I}_2 = 4\underline{I}_1 = 4\times 0.04\angle 30° = (0.16\angle 30°) \text{ A}$$
$$\underline{I}_3 = 2\times\underline{I}_2 = 2\times 0.16\angle 30° = (0.32\angle 30°) \text{ A}$$
$$\underline{V}_{ab} = -6\times\underline{I}_3 = -6\times 0.32\angle 30° = (-1.92\angle 30°) \text{ V}$$

Example 5.4: Matching impedances

An ac power source consists of a voltage source in series with a resistance of 128 Ω. Find the turns ratio of the transformer, that when connected between the power source and the load, would match a load of 8 Ω to the resistance of the power source.

For maximum transfer of power between the power source and the load, the resistance of the source must equal the load resistance referred to the primary. Therefore, if the load resistance is R_L

$$128 = a^2 R_L$$
$$128 = a^2 \times 8$$
$$a = \sqrt{16}$$
$$a = \frac{N_1}{N_2} = 4$$

Chapter 6
Transients

6.1 Introduction

When a switch is closed or opened, in an energized circuit that contains a capacitor or inductor, currents and voltages in different branches take time to change to new steady state values. The time it takes for circuit variables to change, from one set of steady state values to another, is the transient period. Changing currents and voltages during the transient period are referred to as transients. Steady state, or constant, currents and voltages in RL and RC circuits before and after the transient period, were described in Chapter 2. Chapter 6 will describe the methods of mathematically describing circuit variables during the transient period.

Transient currents and voltages in circuits containing a single energy storage element (either an inductor or a capacitor) are determined by solving a first order differential equation. These circuits are referred to as first order transient circuits. The method of solving for transients in first order systems using the differential equation method, and the quicker inspection method, will be described.

An inductor and a capacitor in the same circuit give a second order transient that is described in this chapter using the Laplace transform method.

156

6.2 RL circuit

The switch in the first order RL circuit of Fig. 6.1(i) is open for time $t<0$. No current flows through the inductor during this time. The switch is closed at $t=0$. Fig. 6.1(ii) shows the circuit during the transient period. To determine currents and voltages during the transient the inductor current must first be found. Since the current

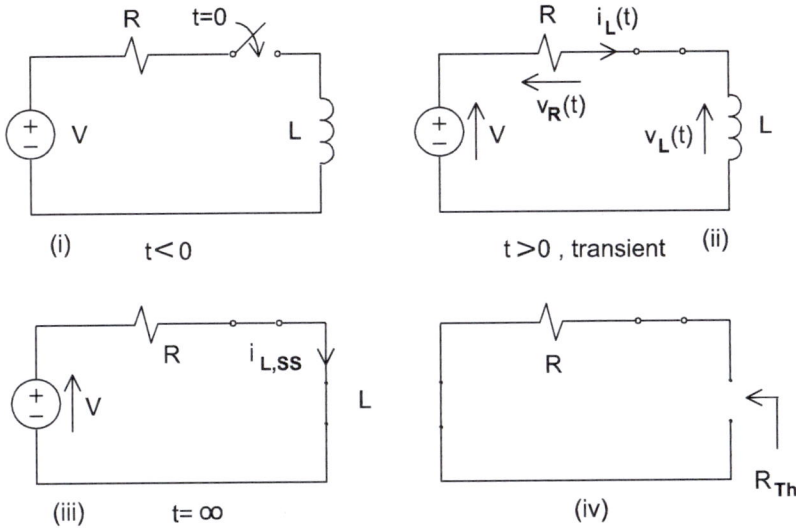

Fig. 6.1(i) RL circuit before the switch is closed (ii) during the transient period (iii) when the circuit has reached the steady state condition and (iv) the Thevenin resistance of the circuit connected to the inductor for $t>0$.

through the inductor cannot change instantaneously, it must be zero just before and just after the switch is closed, at $t = 0^-$ and $t = 0^+$, respectively. This determines the current through the inductor $i_L(0^+)$ at the beginning of the transient as zero. The inductor current and voltage during the transient are $i_L(t)$ and $v_L(t)$, respectively, and $v_R(t)$ is the voltage across the resistor of resistance R. The energy source is a dc voltage of V volts and

157

the inductor has an inductance of L henrys. KVL in the circuit of Fig. 6.1(ii) then gives

$$v_L(t) + v_R(t) = V \tag{6.1}$$

$$L\frac{di_L(t)}{dt} + Ri_L(t) = V \tag{6.2}$$

$$\frac{di_L(t)}{dt} + \left(\frac{1}{L/R}\right)i_L(t) = \frac{V}{L} \tag{6.3}$$

$$\frac{di_L(t)}{dt} + \left(\frac{1}{\tau}\right)i_L(t) = \frac{V}{L} \tag{6.4}$$

The solution to the differential equation of Equ (6.4) is

$$i_L(t) = [i_L(0^+) - i_{L,SS}]e^{-\frac{t}{\tau}} + i_{L,SS} \tag{6.5}$$

Equ (6.5) describes the transient current through the inductor as a function of time where $i_L(0^+)$ is the initial current through the inductor, $i_{L,SS}$ is the final steady state current through the inductor and τ is the time constant of the circuit. The time constant τ is measured in seconds and is given by

$$\tau = \frac{L}{R_{Th}} \tag{6.6}$$

R_{Th}, for a circuit with a single inductor, is the resistance of the circuit looking into the terminals of the inductor during the transient period ($t>0$), after the inductor has been removed and the energy sources have been deactivated (see Fig. 6.1(iv)). R_{Th} is the Thévenin resistance of the circuit connected to the inductor during the transient period. For the present circuit, Fig. 6.1(iv) gives $R_{Th} = R$.

Equ. 6.5 contains a transient and a steady state component and thus describes the inductor current for all time after the switch is closed. The $i_{L,SS}$ term is the final steady state current, while the rest of the right hand side of Equ 6.5 represents the transient response of the circuit.

Let the voltage source provide a voltage of $V=20$ volts, $R=2$ ohms and $L=6$ henrys in the circuit of Fig. 6.1(i). At the end of the transient the circuit reaches a steady state, the inductor current does not change with time, the voltage across it is zero and so the inductor is represented by a short circuit. From KVL the 20 volts of the voltage source then appear across the resistor and from Ohm's law the inductor current is $i_{L,SS}=10$ amperes. The time constant is $\tau = 6/2$ and the variation of the current through the inductor is then given by Equ. 6.5 as

$$i_L(t) = [0-10]e^{-\frac{t}{(6/2)}} + 10 \tag{6.7}$$

$$i_L(t) = 10[1 - e^{-\frac{t}{3}}] \tag{6.8}$$

The transient voltage across the inductor and resistor are now given as

$$v_L(t) = L\frac{di_L(t)}{dt} = 20e^{-\frac{t}{3}} \tag{6.9}$$

$$v_R(t) = Ri_L(t) = 20[1 - e^{-\frac{t}{3}}] \tag{6.10}$$

Equs. 6.8 and 6.9 are plotted in Fig. 6.2.

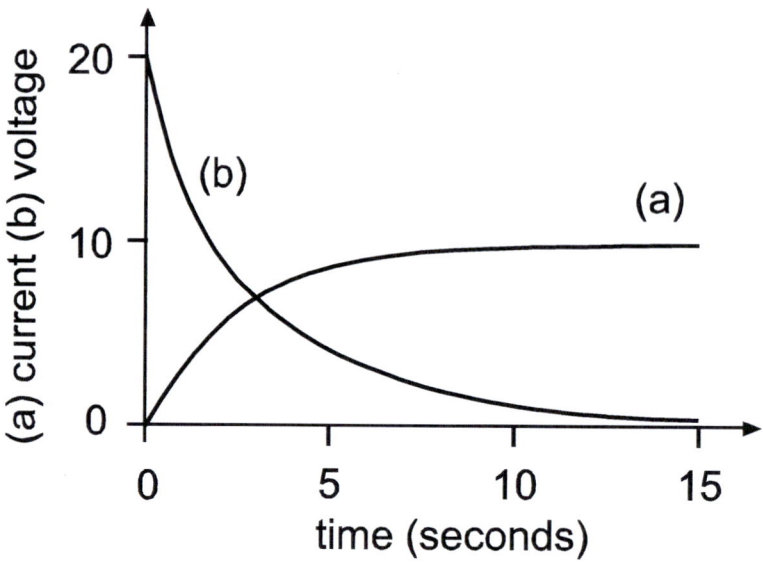

Fig. 6.2 (a) Plot of transient inductor current (Equ.6.8) and (b) transient inductor voltage (Equ. 6.9) for $t>0$, for first order circuit of Fig. 6.1, with a time constant of 3 seconds.

Fig 6.2 shows the variation of the transient current and voltage for the inductor. Both variables have reached 99% of their final steady state values 15 seconds after the switch is closed i.e. after five times the time constant of the circuit. This is applicable to all first order circuits; the transient period is effectively over after five time constants. The time constant helps to determine the duration of the transient.

Fig. 6.2 also shows how the current through the inductor does not change instantaneously at the beginning of the transient but the voltage across the inductor does, going from zero to 20 volts immediately after the switch is closed when the entire supply voltage appears across the inductor. As the current increases with time, the voltage across the resistor starts to increase and at the end of the transient the entire supply voltage appears across the resistor and the inductor is then represented as a short circuit.

160

6.3 RC circuit

The switch in the first order RC circuit of Fig. 6.3(i) is open for time $t<0$. It is assumed that the capacitor is not charged during this time and so the voltage across the capacitor is zero. The switch is closed at $t=0$. Fig. 6.3(ii) shows the circuit during the transient period. To determine currents and voltages during the transient period, the capacitor voltage must first be found. Since the voltage across the capacitor cannot change instantaneously, it must stay zero immediately after the switch is closed, at the beginning of the transient at $t = 0^{+}$.

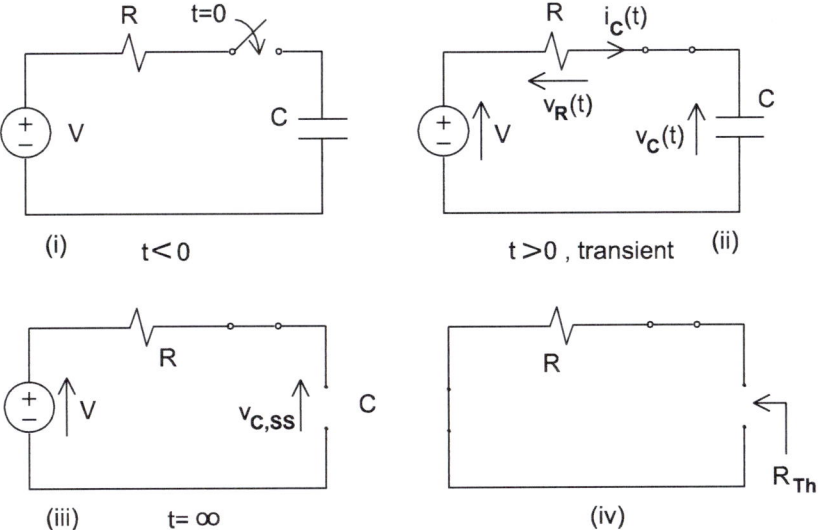

Fig. 6.3(i) RC circuit before the switch is closed (ii) during the transient (iii) when the circuit has reached the steady state condition and (iv) the Thevenin resistance of the circuit connected to the capacitor for $t>0$.

If $v_C(t)$ and $i_C(t)$ are the capacitor voltage and current, respectively, and $v_R(t)$ is the voltage across the resistor R during the transient period shown in Fig. 6.3(ii), then KVL gives

$$V - v_R(t) - v_C(t) = 0 \tag{6.11}$$

$$V - Ri_C(t) - v_C(t) = 0 \tag{6.12}$$

$$V - RC\frac{dv_C(t)}{dt} - v_C(t) = 0 \tag{6.13}$$

$$\frac{dv_C(t)}{dt} + \left(\frac{1}{RC}\right)v_C(t) = \frac{V}{RC} \tag{6.14}$$

$$\frac{dv_C(t)}{dt} + \left(\frac{1}{\tau}\right)v_C(t) = \frac{V}{RC} \tag{6.14}$$

The solution to the above differential equation is

$$v_C(t) = [v_C(0^+) - v_{C,SS}]e^{-\frac{t}{\tau}} + v_{C,SS} \tag{6.15}$$

where $v_C(0^+)$ is the voltage across the capacitor at the start of the transient, $v_{C,SS}$ the final steady state voltage across the capacitor and $\tau = RC$ is the time constant for the circuit. In general, $\tau = R_{Th}C$, where R_{Th} is the Thévenin resistance of the circuit seen by the capacitor during the transient period (see Fig. 6.3(iv)).

Let the supply voltage be $V=20$ volts, $R=4$ ohms and the capacitor have a capacitance of 0.5 farad. Then $\tau = 2$ seconds. At the end of the transient the circuit is in the steady state situation and the capacitor is an open circuit. The constant voltage across the capacitor is then the full 20 volts of the supply voltage. The transient voltages and current are

$$v_C(t) = [0 - 20]e^{-\frac{t}{2}} + 20 \tag{6.16}$$

$$v_C(t) = 20[1 - e^{-\frac{t}{2}}] \tag{6.17}$$

$$i_C(t) = C\frac{dv_C(t)}{dt} = -10(-1/2)e^{-\frac{t}{2}} = 5e^{-\frac{t}{2}} \tag{6.18}$$

$$v_R(t) = Ri_C(t) = 4 \times 5e^{-\frac{t}{2}} = 20e^{-\frac{t}{2}} \tag{6.19}$$

162

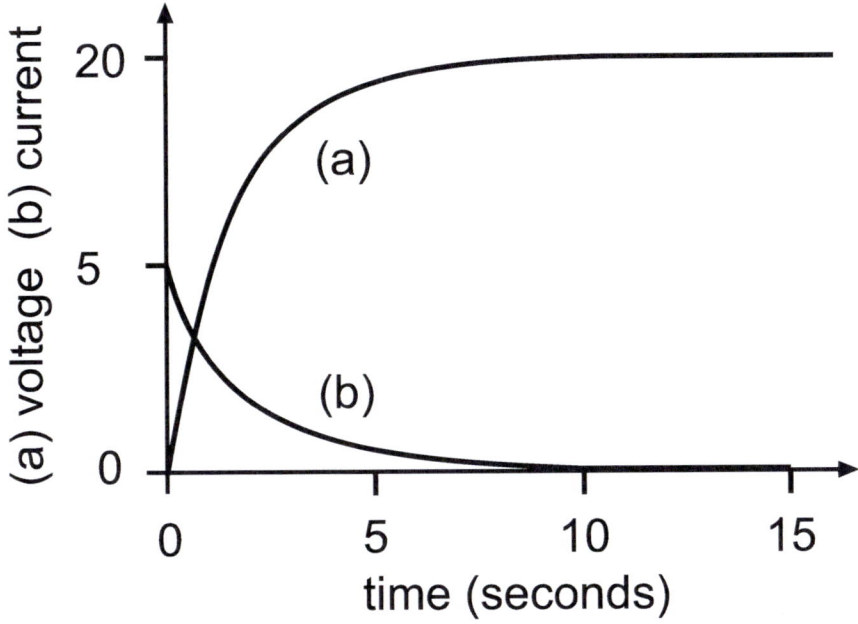

Fig. 6.4(a) Plot of transient capacitor voltage (Equ.6.17) and (b) transient capacitor current (Equ. 6.18) for $t>0$ for RC circuit of Fig. 6.2(ii) with a time constant of 2 seconds.

The transient voltage and current for the capacitor are shown in Fig. 6.4. It shows that the transient lasts for about 10 seconds (5τ). Also, at the beginning of the transient, although the voltage across the capacitor is zero, the capacitor current is at its maximum, so that the rate of flow of charge to the plates of the capacitor is a maximum and the rate of change of the capacitor voltage is a maximum. The maximum charging current is due to the entire supply voltage appearing across the resistor at $t = 0^+$. This charging current diminishes with time, as the voltage across the capacitor increases, and at the end of the transient period, when the current is zero, the capacitor is an open circuit and the voltage across it is the supply voltage of 20 volts.

6.4 The Laplace transform
6.4.1 Definition

A circuit containing a resistor, inductor and capacitor (an RLC circuit) requires the solution of a second order differential equation to describe its transient period. The period is referred to as a second order transient. It is easier to describe the transient period for circuits that contain multiple energy storage elements by using the Laplace transform method. This method transforms the differential equation, describing a voltage or current in the time domain, to an algebraic one which is easier to solve. Taking the inverse Laplace transform then gives the answer as a function of time.

The Laplace transform is defined as

$$\mathcal{L}[f(t)] = F(s) = \int_{0}^{\infty} f(t)e^{-st}dt \qquad (6.20)$$

$\mathcal{L}[f(t)] = F(s)$ is read $F(s)$ is the Laplace transform of $f(t)$. It is assumed that $f(t) = 0$ for $t<0$. The variable t, in Equ. 6.20, is integrated out so that the transform is in terms of s alone. Equ. 6.20 transforms the function $f(t)$ from the time domain, into a function $F(s)$ in the complex frequency domain. Here, s is the complex frequency variable. Also, $s = \sigma + j\omega$ where σ is a real number and ω is the frequency in radians/second. Since e^{-st} is dimensionless, s has the dimension of 1/time, or frequency, and therefore, s is called the complex frequency.

Circuit elements and waveforms representing the excitation applied to the circuit are transformed into the s domain. The laws and methods of circuit analysis are then applied to find an expression for a current or voltage in the s domain. The inverse Laplace transform is then taken, using tabulated Laplace transform pairs, to determine the current or voltage in the time domain.

The inverse Laplace transform is written as

$$\mathcal{L}^{-1}[F(s)] = f(t) \tag{6.21}$$

The inverse Laplace transform is given as

$$f(t) = \frac{1}{\pi} \int_0^\infty \text{Re}\{F(s)e^{st}\}d\omega \tag{6.22}$$

Equ. 6.22 states that signals can be represented by the sum of an infinite number of exponentially rising or decaying complex exponential time waveforms of the form $e^{st} = e^{\sigma t}e^{j\omega t}$, whose strengths (amplitudes) are given by the Laplace transform of the signal.

6.4.2 Transforms of waveforms

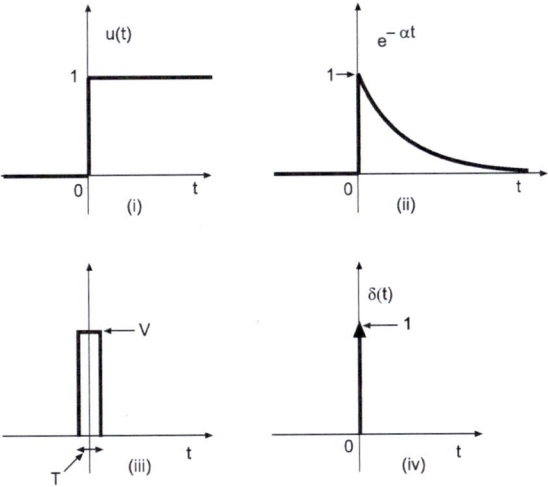

Fig. 6.5 (i) The unit step function (ii) the exponential function (iii) a pulse waveform (iv) the unit impulse (delta) function.

165

The Laplace transforms of some functions are given in Table 6.1. The unit step function, as shown in Fig. 6.5(i), is defined as

$$u(t) = 1 \quad \text{for} \quad t > 0$$
$$= 0 \quad \text{for} \quad t < 0 \tag{6.23}$$

The Laplace transform of the unit step function is

$$\mathcal{L}[u(t)] = \int_0^\infty e^{-st} dt = \left[-\frac{1}{s} e^{-st} \right]_0^\infty \tag{6.24}$$

$$\mathcal{L}[u(t)] = 1/s \tag{6.25}$$

For an exponentially decaying function e^{-at} (see Fig. 6.5(ii)),

$$\mathcal{L}[e^{-at}] = \int_0^\infty e^{-at} e^{-st} dt = \left[-\frac{1}{s+a} e^{-(s+a)t} \right]_0^\infty \tag{6.26}$$

$$\mathcal{L}[e^{-at}] = \frac{1}{s+a} \tag{6.27}$$

The unit impulse, or delta function $\delta(t)$, is defined as

$$\int_{-\infty}^{+\infty} \delta(t) dt = 1 \tag{6.28}$$

$$\delta(t) = 0 \quad \text{for} \quad t \neq 0 \tag{6.29}$$

It is a function which is zero at all times except at $t = 0$ when it has an area of unity. The delta function is the pulse of Fig. 6.5(iii) when $T \rightarrow 0$ and $V \rightarrow \infty$, while the area $VT=1$. It is represented by the vertical arrow of Fig. 6.5(iv) where the arrow points to infinity to indicate that $\delta(t) \rightarrow \infty$ at $t=0$ and that an infinitely narrow pulse

166

is obtained at $t=0$. The height of the arrow gives the area (or amplitude or strength) of the unit impulse function. Taking the Laplace transform of the unit impulse function

$$\mathcal{L}[\delta(t)] = \int_{0^-}^{\infty} \delta(t)e^{-st}\,dt \qquad (6.30)$$

$$\mathcal{L}[\delta(t)] = 1 \qquad (6.31)$$

$$\mathcal{L}^{-1}[1] = \delta(t) \qquad (6.32)$$

$F(s)$	$f(t)$
1	$\delta(t)$
$1/s$	$u(t)$
$1/(s+a)$	e^{-at}
$1/(s+a)^2$	te^{-at}
$1/s^2$	t
$\omega/(s^2+\omega^2)$	$\sin(\omega t)$
$s/(s^2+\omega^2)$	$\cos(\omega t)$
$\omega/[(s+a)^2+\omega^2]$	$e^{-at}\sin(\omega t)$
$(s+a)/[(s+a)^2+\omega^2]$	$e^{-at}\cos(\omega t)$
$1/(s^2+2\alpha s+\omega_0^2)$	$(1/\omega_d)e^{-\alpha t}\sin(\omega_d t)$ $\omega_0 > \alpha,\ \omega_d = \sqrt{\omega_0^2-\alpha^2}$
$s/(s^2+2\alpha s+\omega_0^2)$	$e^{-\alpha t}[\cos(\omega_d t)-(\alpha/\omega_d)\sin(\omega_d t)]$ $\omega_0 > \alpha,\ \omega_d = \sqrt{\omega_0^2-\alpha^2}$

Table 6.1 Laplace transform pairs where $f(t)$ is defined for $t>0$ and $f(t) = 0$ for $t<0$.

6.4.3 Transforms of circuit elements

If $F(s)$ is the Laplace transform of $f(t)$, then the Laplace transform of the derivative of $f(t)$, and the transform of the integral of $f(t)$ are, respectively

$$\mathcal{L}\left[\frac{df(t)}{dt}\right] = sF(s) - f(0^+) \qquad (6.33)$$

$$\mathcal{L}\left[\int_0^t f(t)dt\right] = \frac{1}{s}F(s) \qquad (6.34)$$

Equs. 6.33 and 6.34 are used to find the transformed circuit equivalents of the capacitor and inductor. The transformed circuit equivalents of the resistor, inductor and capacitor follow from the voltage-current relationship for each element so that

$$\mathcal{L}\left[v(t) = Ri(t)\right] \qquad (6.35)$$
$$V(s) = RI(s) \qquad (6.36)$$
$$\mathcal{L}\left[v(t) = L\frac{di(t)}{dt}\right] \qquad (6.37)$$
$$V(s) = sLI(s) - Li(0^+) \qquad (6.38)$$
$$\mathcal{L}\left[v(t) = \frac{1}{C}\int_{-\infty}^t i(t)dt\right] \qquad (6.39)$$
$$V(s) = \frac{1}{sC}I(s) + \frac{v(0^+)}{s} \qquad (6.40)$$

where $i(0^+)$ and $v(0^+)$ are, respectively, the initial current through the inductor and the initial voltage across the capacitor (at the beginning of the transient). The two terms on the right hand side of Equ. 6.38 represent two voltages opposed to one another, one representing the voltage across an inductor of impedance sL and the other a voltage source of $Li(0^+)$. The two terms on the right hand side of Equ. 6.40 represent two voltages both aiding one

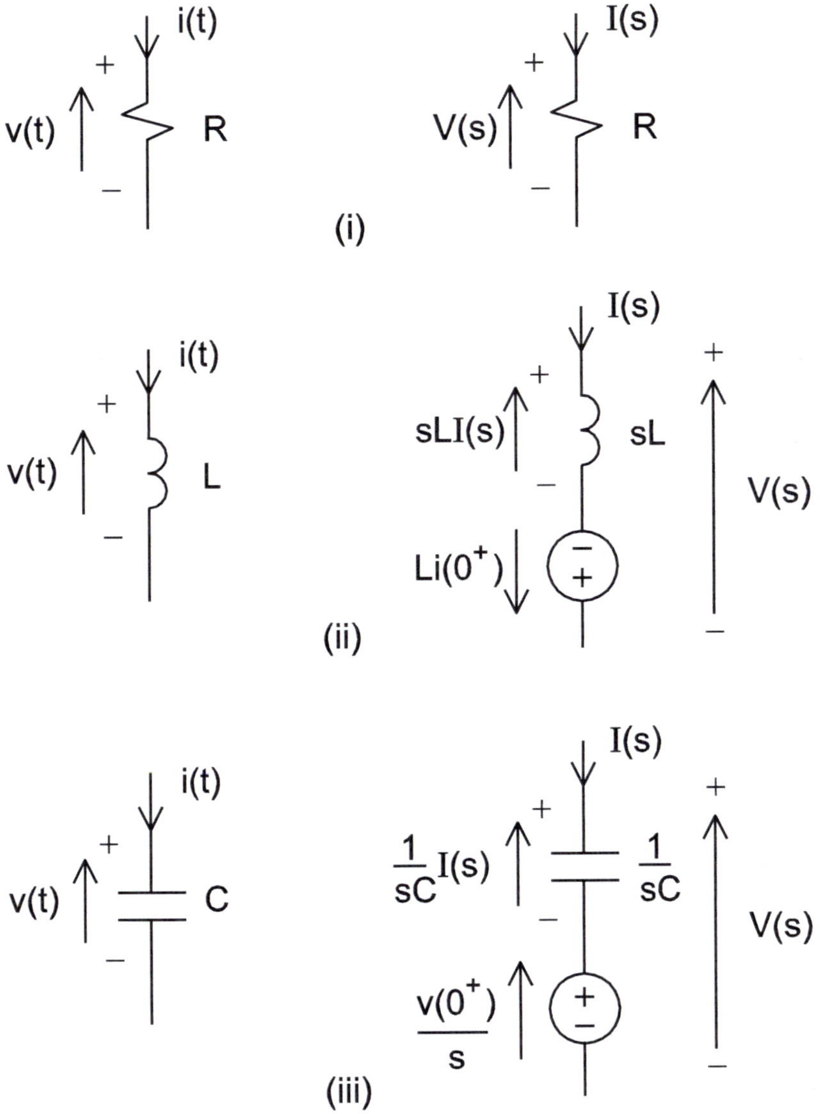

Fig. 6.6 Time and s domain representation for (i) resistor R
(ii) inductor L and (iii) capacitor C.

169

another, with one representing the voltage across a capacitor with impedance $1/(sC)$ and the other a voltage source of $[v(0^+)/s]$. The time and s domain representations of the circuit elements are shown in Fig. 6.6.

If the initial current through the inductor and the initial voltage across the capacitor are zero, Equs. 6.36, 6.38 and 6.40 give the voltage-current equations for R, L and C in the s domain as

$$V(s) = RI(s) \tag{6.41}$$
$$V(s) = sLI(s) \tag{6.42}$$
$$V(s) = \frac{1}{sC} I(s) \tag{6.43}$$

Impedance in the s domain is defining as

$$Z(s) = \frac{V(s)}{I(s)} \tag{6.44}$$

Then the resistor, inductor and capacitor, of resistance R, inductance L and capacitance C, respectively, are represented in the s domain by impedances R, sL and $(1/sC)$. Voltage and current sources with time waveforms $v(t)$ and $i(t)$ are represented in the s domain by the Laplace transforms $V(s)$ and $I(s)$, of waveforms $v(t)$ and $i(t)$, respectively. The laws and principles discussed in Chapter 1 are applied to solve for a branch current or voltage in the transformed circuit in the s domain. In the case of a second order circuit, this leads to an expression of the form $s^2 + 2\alpha s + \omega_0^2$ in the denominator. If $\alpha > \omega_0$ the quadratic has real and distinct roots and the circuit is said to be overdamped. If $\alpha < \omega_0$ the roots are complex and the system is underdamped. When $\alpha = \omega_0$, the circuit is critically damped. The time variation of the circuit parameter is then found by inverse transforming the expression using the Laplace transform pairs of Table 6.1.

Example 6.1: Transient in RL circuit

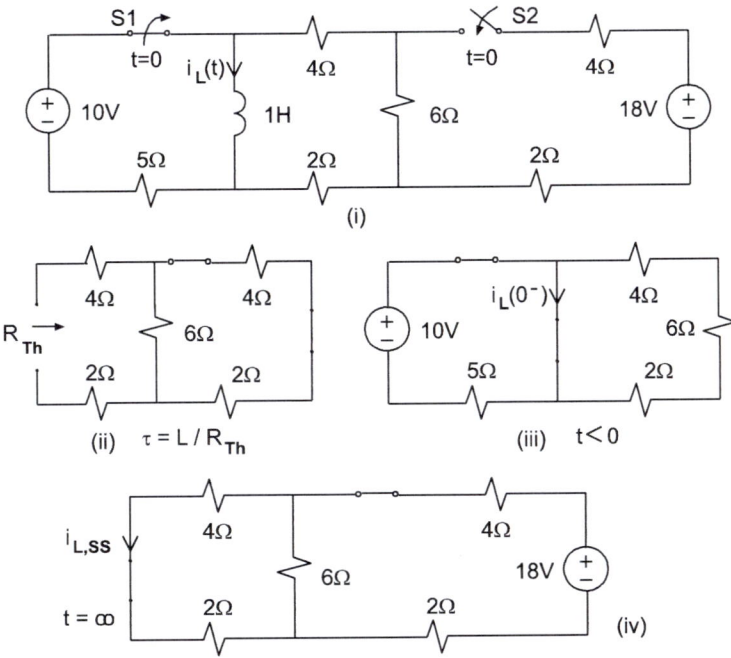

Fig. 6.7(i) Circuit for Example 6.1.

The circuit of Fig. 6.7(i) is in the steady state for $t<0$. Switch S1 is opened and S2 closed at $t=0$. Find the transient current $i_L(t)$ for $t>0$.

$R_{Th} = 4 + 6 \| 6 + 2 = 9\Omega$. (Fig. 6.7(ii)).

For $t<0$, $i_L(0^-) = 10/5 = 2\,\text{A} = i_L(0^+)$ (From Fig. 6.7(iii)).

Ohm's law and current division in the circuit of Fig. 6.7(iv) gives, at $t=\infty$, $i_{L,SS} = 1/2[18/(4+3+2)] = 1$ A. Then the transient is

$i_L(t) = [i_L(0^+) - i_{L,SS}]e^{-t/(L/R_{Th})} + i_{L,SS}$

$i_L(t) = [2-1]e^{-t/(1/9)} + 1$

$i_L(t) = (e^{-9t} + 1)$ A.

Example 6.2: Transient in RC circuit

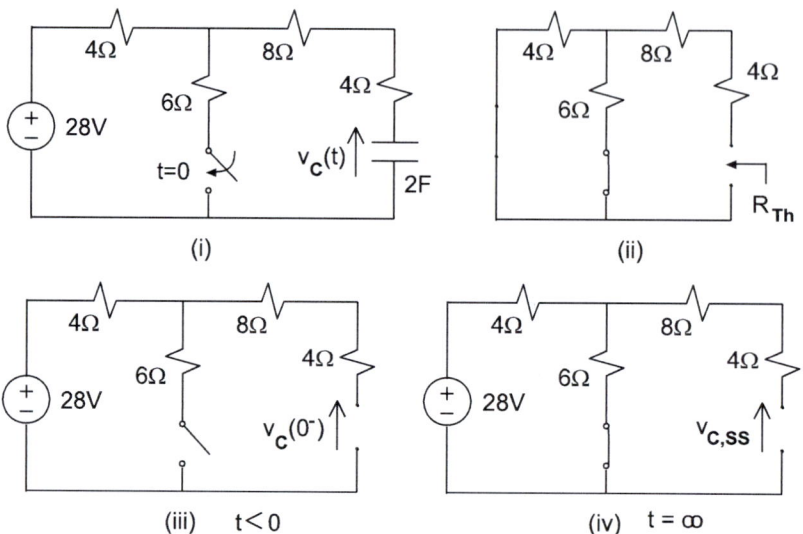

Fig. 6.8(i) Circuit for Example 6.2.

The circuit of Fig. 6.8(i) is in the steady state for $t<0$. The switch is closed at $t = 0$. Find the transient voltage $v_C(t)$ for $t>0$ and the voltage across the capacitor 5 seconds after the switch is closed.

The resistance connected to the capacitor terminals in the transient period (Fig. 6.8(ii)) is $R_{Th} = 4 + 8 + 6 \| 4 = 12 + 2.4 = 14.4\,\Omega$.

From Fig. 6.8(iii) for $t<0$, KVL gives $v_C(0^-) = 28\,\text{V} = v_C(0^+)$.

From the circuit of Fig. 6.8(iv) for $t = \infty$, voltage division gives
$v_{C,SS} = [6/(6+4)]28 = 16.8\,\text{V}$.

$v_C(t) = [v_C(0^+) - v_{C,SS}]e^{-t/(R_{Th}C)} + v_{C,SS}$

$v_C(t) = [28 - 16.8]e^{-t/(14.4 \times 2)} + 16.8$

$v_C(t) = 11.2(e^{-t/28.8} + 1.5)\,\text{V}$.

$v_C(5) = 11.2(e^{-5/28.8} + 1.5) = 26.215\ \text{V}$.

172

Example 6.3: Transient in RLC circuit

A step function of 10V is applied at time t=0 in the RLC circuit of Fig. 6.9(i). Find $i(t)$ and $v(t)$ for $t>0$ for R=1Ω, 2Ω and 4Ω.

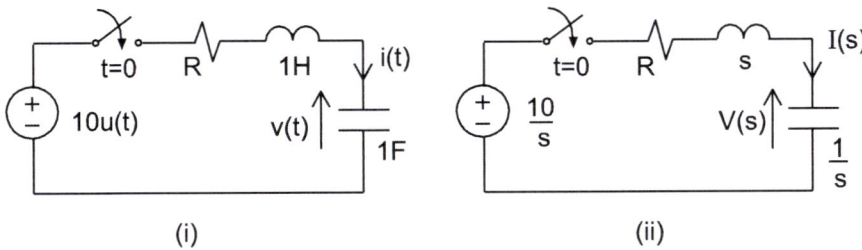

(i)   (ii)

Fig. 6.9 Circuit for Example 6.3 in (i) the time domain and (ii) the
s domain.

$i(t)$

KVL in the transformed circuit of Fig. 6.9(ii) gives

$$I(s) = \frac{10/s}{R + sL + 1/sC} = \frac{10}{s^2 + Rs + 1}$$

a) For R=1Ω

$$I(s) = \frac{10}{s^2 + s + 1}, \quad \alpha = 0.5, \omega_0 = 1, \alpha < \omega_0, \omega_d = 0.87.$$

The circuit is underdamped and Table 6.1 gives

$$i(t) = \mathcal{L}^{-1}\left[\frac{10}{(s^2 + s + 1)}\right] = \left(\frac{10}{0.87}\right)e^{-t/2}\sin(0.87t)$$

$$i(t) = 11.55e^{-t/2}\sin(0.87t) \text{ A}$$

b) For _R=2Ω_

$$I(s) = \frac{10}{s^2 + 2s + 1} = \frac{10}{(s+1)^2}, \quad \alpha = \omega_0 = 1.$$

The circuit is critically damped and Table 6.1 gives

$$i(t) = \mathcal{L}^{-1}\left[\frac{10}{(s+1)^2}\right]$$

$$i(t) = 10te^{-t} \text{ A}$$

c) For _R=4Ω_

$$I(s) = \frac{10}{s^2 + 4s + 1}$$

$\alpha = 2$, $\omega_0 = 1$. $\alpha > \omega_0$ and the circuit is overdamped. Expanding by calculator and using Table 6.1 gives

$$I(s) = \frac{2.89}{s+0.27} - \frac{2.89}{s+3.73}$$

$$\mathcal{L}^{-1}[I(s)] = \mathcal{L}^{-1}\left[\frac{2.89}{s+0.27}\right] - \mathcal{L}^{-1}\left[\frac{2.89}{s+3.73}\right]$$

$$i(t) = 2.89(e^{-0.27t} - e^{-3.73t}) \text{ A}$$

Fig. 6.10 shows $i(t)$ for the three cases. After the fluctuations of the current in the transient period are over, the current settles to the final steady state value, which in this case is zero. The solution using the Laplace transform method gives both the transient and the steady state current.

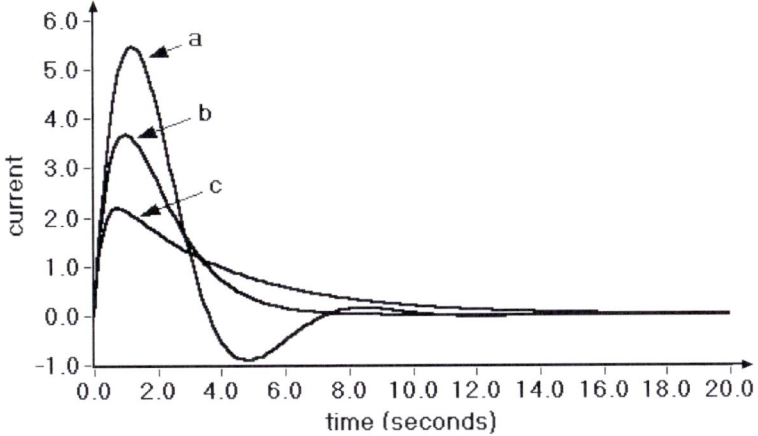

Fig. 6.10 Plot of $i(t)$ for Example 6.3 for a) underdamped
b) critically damped and c) overdamped circuit.

$v(t)$

Voltage division in the circuit of 6.9(ii) gives

$$V(s) = \left(\frac{1/sC}{R + sL + 1/sC} \right) \frac{10}{s} = \left(\frac{1/s}{R + s + 1/s} \right) \frac{10}{s}$$

a) For $R=1\Omega$

$$V(s) = \frac{10}{s^3 + s^2 + s}$$

Expanding the right hand side of the above equation by calculator gives

$$V(s) = \frac{10}{s} - \frac{10}{s^2 + s + 1} - \frac{10s}{s^2 + s + 1}$$

175

Taking the inverse Laplace transform from Table 6.1 gives

$$v(t) = [10 - (10/0.87)e^{-t/2} \sin(0.87t)...$$
$$... - 10e^{-t/2}[\cos(0.87t) - (0.5/0.87)\sin(0.87t)]$$
$$v(t) = \{10 - 11.49e^{-t/2} \sin(0.87t)...$$
$$... - 10e^{-t/2}[\cos(0.87t) - 0.57\sin(0.87t)]\}$$
$$v(t) = 10\{1 - 1.149e^{-t/2} \sin(0.87t)...$$
$$... - e^{-t/2}[\cos(0.87t) - 0.575\sin(0.87t)]\}$$

b) For $R=2\Omega$

$$V(s) = \frac{10}{s^3 + 2s^2 + s}$$
$$V(s) = \frac{10}{s} - \frac{10}{s+1} - \frac{10}{(s+1)^2}$$
$$v(t) = 10(1 - e^{-t} - te^{-t}) \text{ V}$$

c) For $R=4\Omega$

$$V(s) = \frac{10}{s^3 + 4s^2 + s}$$
$$V(s) = \frac{10}{s} + \frac{0.77}{s+3.73} - \frac{10.77}{s+0.27}$$
$$v(t) = 10 + 0.77e^{-3.73t} - 10.77e^{-0.27t}$$
$$v(t) = 10(1 + 0.077e^{-3.73t} - 1.077e^{-0.27t}) \text{ V}$$

Fig. 6.11 shows $v(t)$ for the three cases. At the end of the transient period the voltage across the capacitor reaches the steady state value of 10 volts.

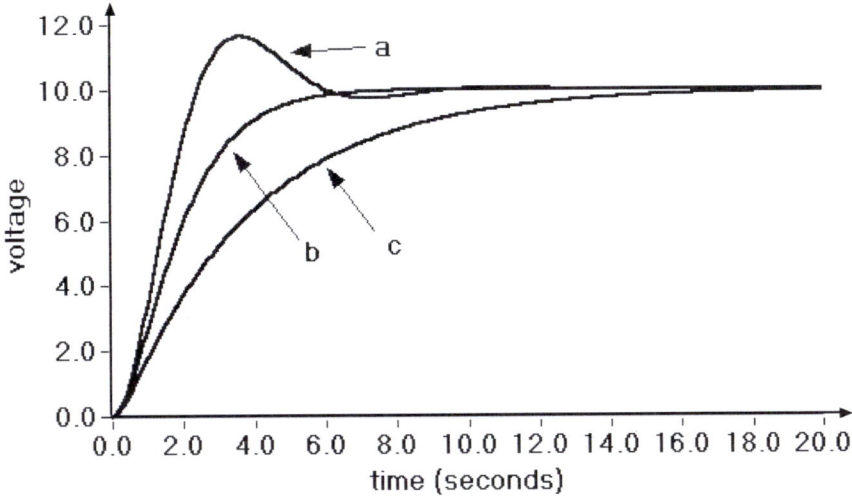

Fig. 6.11 Plot of $v(t)$ for Example 6.3 for a) underdamped
b) critically damped and c) overdamped circuit.

Appendix 6.A: Derivation of last two transforms of Table 6.1.

$$\mathcal{L}^{-1}[1/(s^2 + 2\alpha s + \omega_0^2)]$$

$$= \mathcal{L}^{-1}\{[1/(\sqrt{\omega_0^2 - \alpha^2})]\}\{[\sqrt{(\omega_0^2 - \alpha^2)}]/[(s+\alpha)^2 + (\omega_0^2 - \alpha^2)]\}$$

$$= \mathcal{L}^{-1}\{1/\omega_d\}\{\omega_d/[(s+\alpha)^2 + \omega_d^2]\} = (1/\omega_d)e^{-\alpha t}\sin(\omega_d t) \quad (A1)$$

The eighth transform pair of Table 6.1 was used in the derivation
of Equ. A1.

$$\mathcal{L}^{-1}[s/(s^2 + 2\alpha s + \omega_0^2)]$$

$$= \mathcal{L}^{-1}\left\{\frac{s+\alpha}{(s+\alpha)^2 + (\omega_0^2 - \alpha^2)} - \left(\frac{\alpha}{\sqrt{\omega_0^2 - \alpha^2}}\right)\left[\frac{\sqrt{\omega_0^2 - \alpha^2}}{(s+\alpha)^2 + (\omega_0^2 - \alpha^2)}\right]\right\}$$

177

$$= \mathcal{L}^{-1} \{ [\frac{s+\alpha}{(s+\alpha)^2 + \omega_d^2}] - (\frac{\alpha}{\omega_d})[\frac{\omega_d}{(s+\alpha)^2 + \omega_d^2}] \}$$

$$= e^{-\alpha t} \cos(\omega_d t) - (\alpha / \omega_d) e^{-\alpha t} \sin(\omega_d t) \qquad (A2)$$

The eighth and ninth transform pairs of Table 6.1 were used in the derivation of Equ. A2.

Equations A1 and A2 describe the response of underdamped systems ($\alpha < \omega_0$), where

$\omega_d = \sqrt{\omega_0^2 - \alpha^2}$ = damped natural frequency

ω_0 = undamped natural frequency

α^{-1} = time constant of the time waveform

$T = 2\pi / \omega_d$ = period of the time waveform.

Chapter 7
Frequency response

7.1 Introduction

In the ac analysis of circuits in Chapter 3, the source of excitation was a sinusoid of a single frequency. If the input signal is an arbitrary waveform, however, it can be represented by the sum of sinusoids of different frequencies. If the voltage appearing across a certain branch of the circuit is of interest, the frequency response of the circuit determines which frequencies present in the input signal appear in the branch voltage. In a circuit that functions as a filter, certain frequencies present in the input signal are allowed to pass to the output, while others are suppressed, or attenuated. A frequency that passes to the output is represented by a sinusoid that has the same amplitude at both the input and output. A frequency that is suppressed is a sinusoid whose amplitude at the output is reduced (attenuated) compared to its amplitude at the input. The frequency selection is decided by the frequency response of the filter.

The frequency response is determined by the transfer function. The transfer function is the ratio of the output variable to the input variable. If the circuit is excited by an input voltage of variable frequencies of the same amplitude, then the transfer function determines the frequency and phase of sinusoids that make up the output voltage. The frequency response of a first order lowpass and a first order highpass filter will be determined in this chapter.

7.2 Lowpass filter

A lowpass filter is shown in the circuit of Fig. 7.1. The angular frequency ω of the input voltage $V_i(\omega)$ is variable. The voltage $V_0(\omega)$, appearing across the capacitor, is the output voltage.

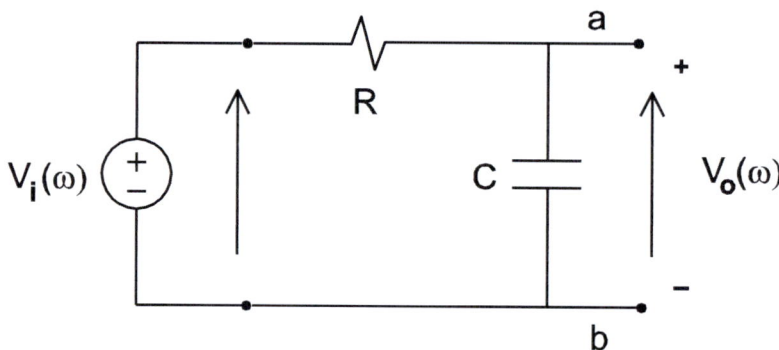

Fig. 7.1 A lowpass filter.

The impedance of the capacitor is $1/j\omega C$. At low frequencies the capacitor is an open circuit and the input voltage will appear across the capacitor as the output voltage. Therefore, low frequencies of the input voltage will be allowed to pass to the output. At high frequencies, the capacitor impedance will be small; ultimately the capacitor will be a short circuit, so that terminals *a* and *b* are at the same potential, and the output voltage will then be zero. Therefore, high frequencies at the input will not appear at the output. The circuit allows low frequency signals at the input to pass to the output, while preventing high frequencies from doing so.

Treating the input and output voltages as phasors with variable frequency, the relationship between them is given by the voltage division rule

$$V_0 = \left(\frac{1/j\omega C}{R + 1/j\omega C} \right) V_i \qquad (7.1)$$

The transfer function of the filter is then written as

$$\underline{H} = H\angle\theta = \frac{V_0}{V_i} = \frac{1/j\omega C}{R + 1/j\omega C} = \frac{1}{1 + j\omega RC} = \frac{1}{1 + j(\omega/\omega_c)} \qquad (7.2)$$

where $\omega_c = 1/RC$. The transfer function is a complex quantity and its magnitude and phase angle are, respectively,

$$|\underline{H}| = |H\angle\theta| = \left|\frac{V_0}{V_i}\right| = \frac{|V_0\angle\theta_0|}{|V_i\angle\theta_i|} = \frac{1}{\sqrt{1 + (\omega/\omega_c)^2}} \qquad (7.3)$$

$$\theta = \theta_0 - \theta_i = -\tan^{-1}(\omega/\omega_c) \qquad (7.4)$$

The magnitude H of the transfer function is plotted against the logarithm of the normalized frequency ω/ω_c in Fig. 7.2(i). For low frequencies the magnitude of the transfer function is flat and has the value of one. Therefore, low frequency signals applied to the input pass to the output with no change in their amplitude. The magnitude of the transfer function begins to decrease as the normalized frequency increases above one (where the angular frequency $\omega = \omega_c = 1/RC$). The angular frequency ω_c is the cutoff frequency, or the half power or the 3dB frequency, of the filter. At ω_c the magnitude of the transfer function is $1/\sqrt{2}$, or 0.707. At ω_c the average power delivered to a load by the circuit is a half the power delivered at dc. The power delivered at ω_c is $10\log_{10}(0.707^2/1) = 3\,\mathrm{dB}$ below that at dc (dB= decibels).

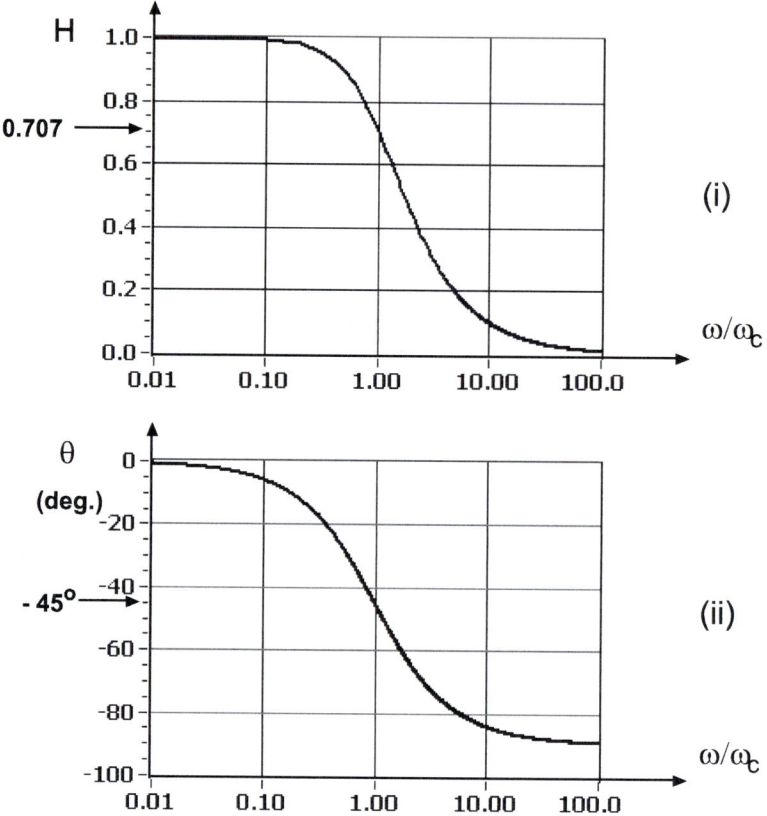

Fig. 7.2 (i) Magnitude H and (ii) phase angle θ, as a function of normalized frequency ω/ω_c, of the transfer function of the lowpass filter of Fig. 7.1.

Input frequencies below ω_c (low frequencies) pass through to the output with little or no decrease in amplitude, while those above ω_c (high frequencies) are attenuated. The circuit is therefore called a lowpass filter. The frequency range below ω_c is the passband

and that above ω_c the stopband. The choice of values of R and C decides the width of the passband.

The phase angle of the transfer function θ is plotted as a function of the logarithm of the normalized frequency in Fig. 7.2(ii). θ is also the difference between the phase angle, θ_0, of the output voltage, and the phase angle, θ_i, of the input voltage. Fig. 7.2(ii) shows that the phase angle difference between the input and output signals is small, at low frequencies, but that it increases with increase in frequency. The magnitude of the phase angle difference is below 45° in the passband and above 45° in the stopband.

7.3 Highpass filter

The circuit for a highpass filter is shown in Fig. 7.3. $V_i(\omega)$ is the input signal.

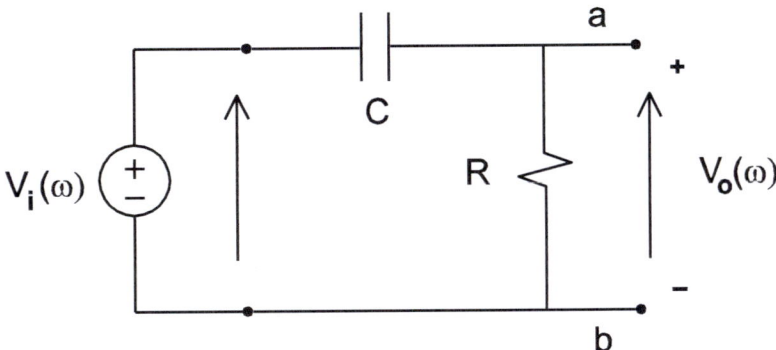

Fig. 7.3 A highpass filter.

The voltage across the resistor is the output signal, $V_0(\omega)$, and voltage division gives it as

$$V_0 = \left(\frac{R}{R + 1/j\omega C} \right) V_i \tag{7.5}$$

183

The transfer function of the filter is

$$\underline{H} = H\angle\theta = \frac{V_0}{V_i} = \frac{1}{1 + 1/j\omega RC} = \frac{1}{1 - j(\omega_c/\omega)} \qquad (7.6)$$

where $\omega_c = 1/RC$. The transfer function is a complex quantity and its magnitude and phase angle are, respectively,

$$|\underline{H}| = |H\angle\theta| = \left|\frac{V_0}{\underline{V}_i}\right| = \frac{|V_0\angle\theta_0|}{|V_i\angle\theta_i|} = \frac{1}{\sqrt{1 + (\omega_c/\omega)^2}} \qquad (7.7)$$

$$\theta = \theta_0 - \theta_i = \tan^{-1}(\omega_c/\omega) \qquad (7.8)$$

The magnitude, H, of the transfer function is plotted against the logarithm of the normalized frequency, ω/ω_c, in Fig. 7.4(i). The breakpoint frequency is at $\omega = \omega_c = 1/RC$. Sinusoids at the input with frequencies higher than the breakpoint frequency pass to the output, while those below ω_c do not. The circuit therefore acts as a highpass filter. The range of frequencies above ω_c is the passband while the stopband lies below ω_c.

The phase angle θ, of the transfer function, is plotted as a function of the logarithm of the normalized frequency in Fig. 7.4(ii). It shows that the phase angle difference between input and output signals is small at high frequencies, but that it increases with decrease in frequency. The phase angle difference is below 45° in the passband and above 45° in the stopband.

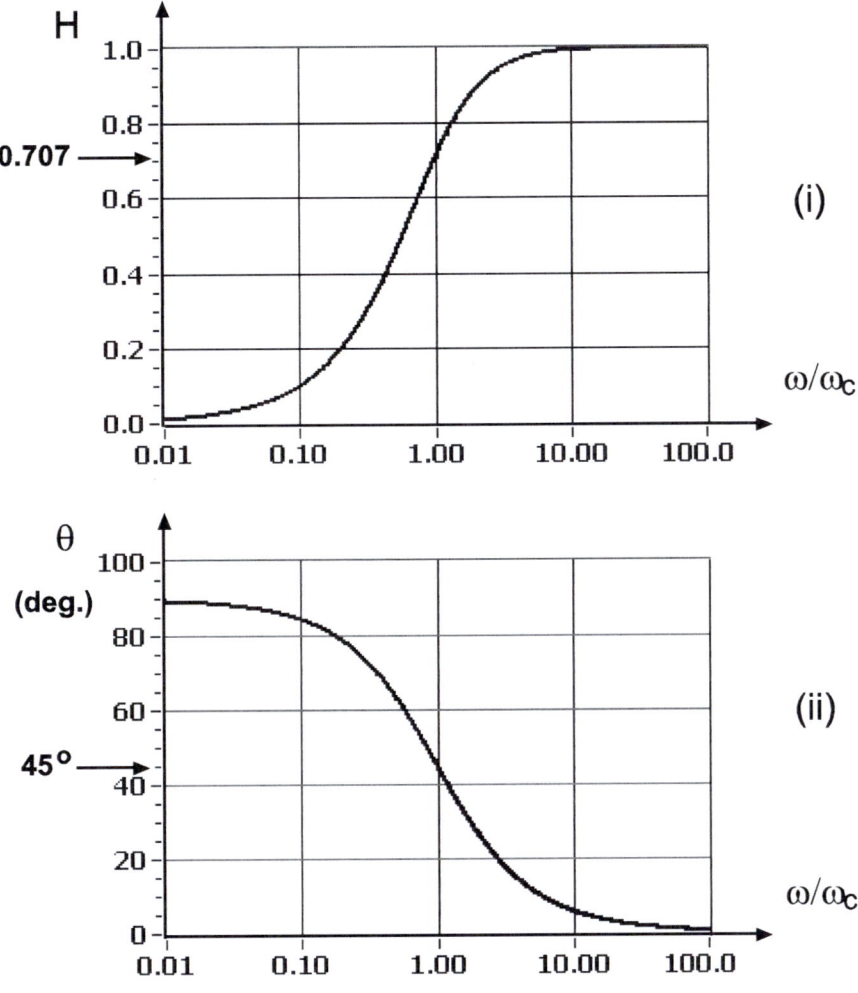

Fig. 7.4 (i) Magnitude H and (ii) phase angle θ, as a function of normalized frequency ω / ω_c, of the transfer function of the highpass filter of Fig. 7.3.

Chapter 8
The operational amplifier

8.1 Introduction

Operational amplifiers, or op amps, are available as integrated circuits (ICs). They contain transistors and resistors built into a single crystal of silicon. The IC is encapsulated for protection and is available as a dual-in-line package (DIP) with electrically conducting pins for insertion into a printed circuit board. With external components and dc voltage sources connected to different pin contacts, the op amp is used as a voltage amplifier or to perform mathematical operations, such as addition, subtraction, differentiation and integration (hence its name). The electrical properties of the op amp will be discussed in this chapter.

8.2 Properties of op amp

   The circuit of Fig. 8.1 shows the connections to the popular 741 op amp. The rectangular outline represents the edges of the DIP where the pin numbers appear on either side. The triangle is the circuit symbol of the op amp. The minus (-) and the plus (+) in the triangle are the inverting and non-inverting terminals of the op amp

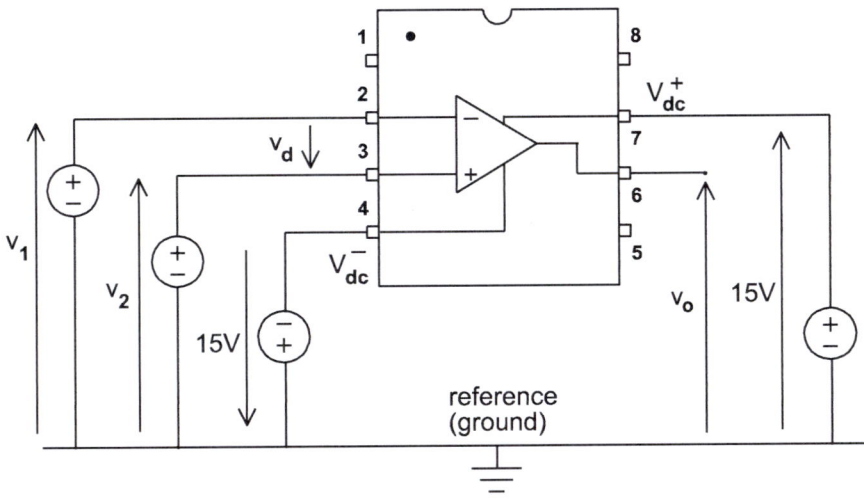

Fig. 8.1 Electrical connections to the 741 op amp.

and these terminals are connected internally to pins 2 and 3 of the DIP. Input voltages v_1 and v_2 are applied externally to the inverting and non-inverting terminals, respectively. Both v_1 and v_2 are with respect to ground (0 potential) and can be dc or ac. The voltage across the input terminals $v_d = v_2 - v_1$ is called the differential input of the op amp. The output v_o is the voltage of pin 6 with respect to ground.

   Electrical power is needed by the op amp for its internal circuit to function properly. A dc voltage of V_{dc}^+ is applied to pin 7 so that this contact is positive with respect to ground, while a dc voltage

187

of V_{dc}^- is applied to pin 4 so that this pin is negative with respect to the reference potential. The magnitude of the dc supply voltages is typically 9V to 15V and is shown as 15V in Fig. 8.1.

It is more usual not to show the dc supply and to use the circuit symbol of the op amp of Fig. 8.2.

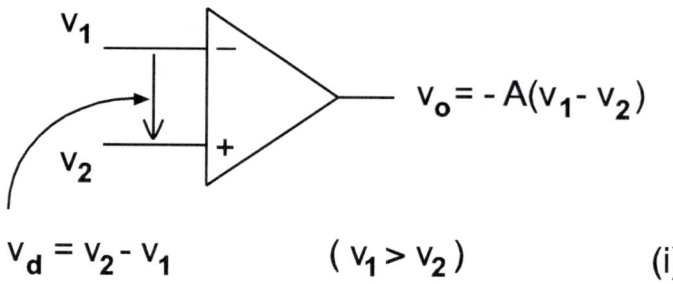

$$v_o = -A(v_1 - v_2)$$

$$v_d = v_2 - v_1 \qquad (v_1 > v_2) \qquad \text{(i)}$$

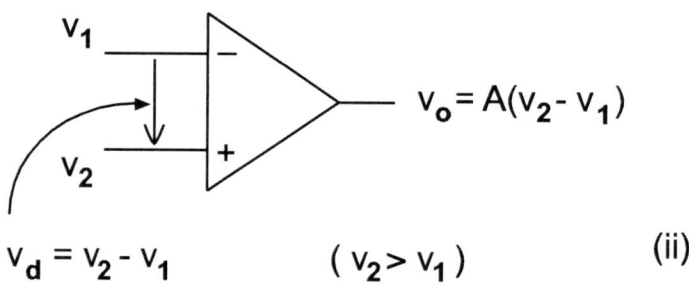

$$v_o = A(v_2 - v_1)$$

$$v_d = v_2 - v_1 \qquad (v_2 > v_1) \qquad \text{(ii)}$$

Fig. 8.2 The symbol of the op amp with applied input voltages v_1 and v_2 for the case when (i) the inverting input is at a higher potential than the non-inverting input and (ii) the non-inverting input is at a higher potential than the inverting terminal.

In Fig. 8.2 v_d is the voltage of the non-inverting terminal with respect to the inverting one. If, as in Fig. 8.2(i), $v_d < 0$ so that the inverting input is at a higher potential than the non-inverting one, then a negative voltage v_o, of magnitude A times the differential input, appears at the output. If, as in Fig. 8.2(ii), $v_d > 0$ so that the non-inverting input is at a higher potential than the inverting one, then a positive voltage v_o, of magnitude A times the differential input, appears at the output.

Therefore, if the inverting input is more positive than the non-inverting one then the output is negative. If the non-inverting input is more positive than the inverting one then the output is positive; hence the names of the two input terminals.

A is the open circuit voltage gain of the op amp. A is high, of the order of 10^5, so that the differential input v_d is greatly amplified at the output.

An electrical model of an actual op amp is shown in Fig. 8.3. R_i

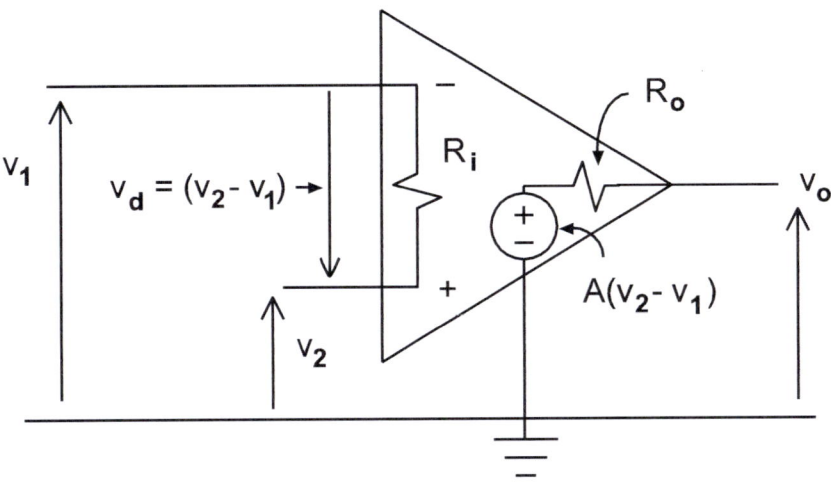

Fig 8.3 A model of the op amp.

is the Thévenin equivalent resistance looking into the input terminals or simply the input resistance of the op amp. R_0 is the Thévenin equivalent resistance looking into the output terminals (the output resistance). The voltage source at the output is a voltage-controlled source, since its value is decided by the

189

differential input v_d. R_i is high; of the order of $10 \times 10^6 \Omega$ (10 megaohms, 10MΩ). R_0 Is small (of the order of 10Ω) and is usually ignored. The output stage of the circuit then shows that the differential input v_d results in an output voltage of Av_d.

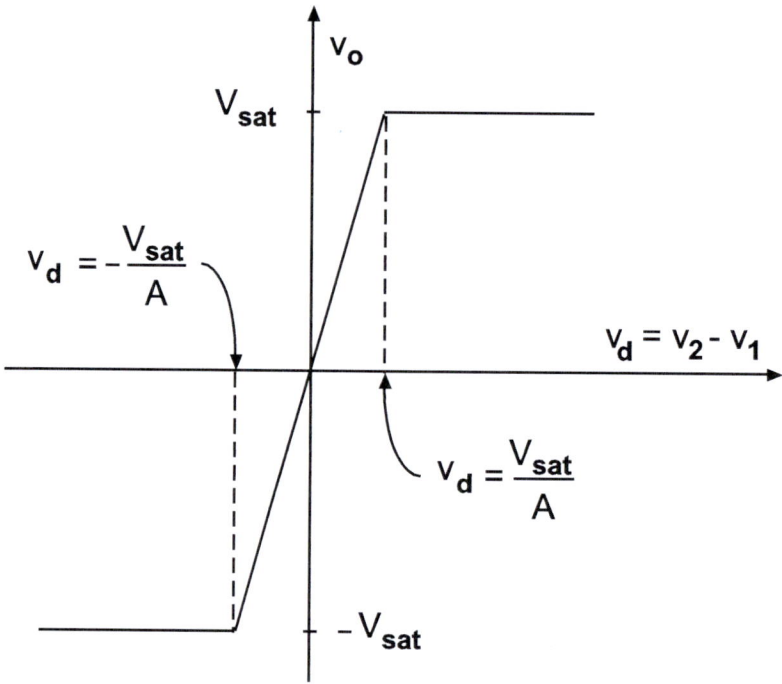

Fig. 8.4 The input-output relationship of the op amp.

Fig. 8.4 shows the input-output relationship of the op amp. The input is the differential voltage v_d and the output is v_o. When $v_2 > v_1$, $v_o > 0$, and when $v_1 > v_2$, $v_o < 0$. For small, positive values of v_d, the output is positive and it varies linearly with the input. The slope of the straight line is very large (it equals A) and so a small increase in the input results in a large increase in the output. The output eventually reaches the maximum of V_{sat}, the saturation voltage, which is close to, but slightly less than, the supply voltage

V_{dc}^{+}. Saturation starts at $v_d = V_{sat}/A$. Further increase in v_d results in no further increase in v_o. Similarly, the variation between the input voltage and the output one is linear for small negative values of v_d, with the negative output eventually saturating at $-V_{sat}$. The region where $-V_{sat} \le v_0 \le V_{sat}$ and $(-V_{sat}/A) \le v_d \le (V_{sat}/A)$, is called the linear region of operation of the op amp. For $v_d \ge (V_{sat}/A)$ the op amp is in the positive saturation region, and for $v_d \le (-V_{sat}/A)$ it is in the negative saturation region.

Since A is very large the op amp saturates at small values of v_d. For instance, if V_{sat}=10V and A=10^5, the maximum value of v_d in the linear region is $10/10^5 = 100 \times 10^{-6}$ V or 100 µV (1 µV=1 microvolt=1×10^{-6} volt or a millionth of a volt). This means that in the linear region the two input terminals of the op amp are approximately at the same potential. This leads to the ideal model of the op amp that is shown in Fig. 8.5.

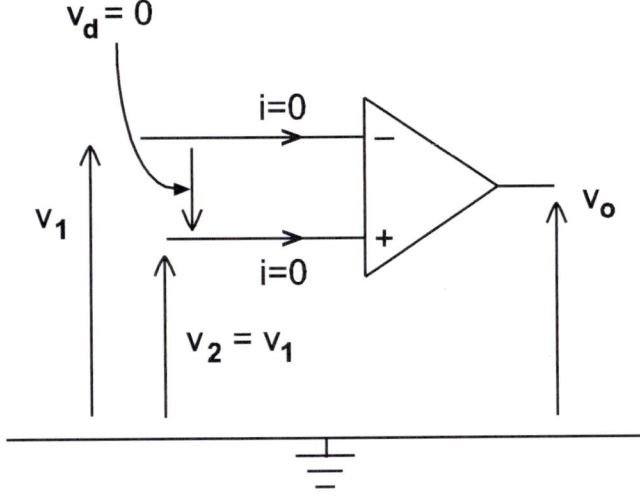

Fig. 8.5 The ideal op amp.

The two inputs of the ideal op amp are assumed to be at the same potential and are said to form a virtual short circuit. The word virtual is used to describe this condition to distinguish it from a short circuit. A short circuit between two terminals implies zero resistance between the terminals and a current can flow between them. In the virtual short circuit, however, no current flows between the inverting and non-inverting terminals. This is because of the very high resistance between the terminals. This is the input resistance of the op amp, which for the ideal op amp, is assumed to be infinite.

The ideal op amp is a good approximation of the actual one. It can be used to design and analyze circuits involving actual op amps. It assumes that the inputs of the op amp are at the same potential, and that no current can flow into the inverting or the non inverting terminals. These properties of the ideal op amp are shown in Fig. 8.5.

The op amps considered in this chapter are all assumed to be operating in the linear region. If the signal to be amplified is very small, then it can be applied as a differential input to the op amp, as in Figs. 8.2 and 8.3, and the amplified signal then appears at the output. In this case, there is no need to ignore the small difference between the input terminals, which is the signal of interest. Thermocouple, strain gauge and physiological signals can be amplified in this way, although care should be taken so that the amplifier does not saturate. The assumptions about the ideal op amp of Fig. 8.5 are useful in analyzing op amp circuits that involve feedback between the output and input, through external components, as seen in the following examples. In such circuits the input voltage can be considerably larger than the 100 μV differential input mentioned earlier, without however, the amplifier saturating.

Example 8.1: Inverting amplifier

Find the output voltage v_o, in terms of the input voltage v_1 and the external resistors.

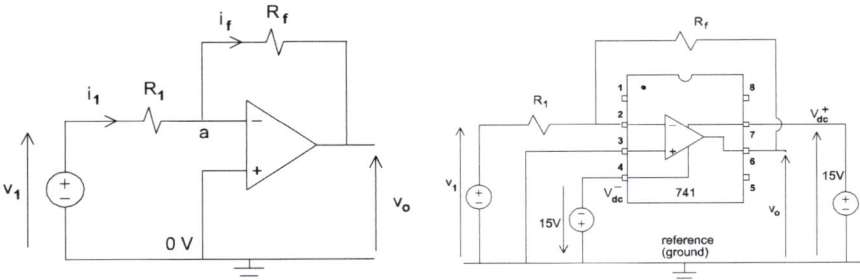

Fig. 8.6 Circuit for Example 8.1.

Since no current flows into the inverting terminal, by KCL at node a, the current through R_1, i_1, equals the feedback current i_f. Also, since the inverting and non-inverting inputs of the op amp are at the same potential, and the non-inverting input is at ground potential, then node a is also at zero potential. Then KCL and Ohm's law give

$$\frac{v_1 - 0}{R_1} = \frac{0 - v_0}{R_f}$$

$$v_0 = -\frac{R_f}{R_1} v_1$$

The minus sign indicates that the polarity of the output voltage is opposite to that of the input. For $R_1 = 1\,M\Omega$, $R_f = 3\,M\Omega$ and a supply voltage of $\pm 15V$, a dc input of 2V results in $v_0 = -6\,V$. Since $|v_0| < 15V$, the op amp is in the linear region of operation.

193

Example 8.2: Non-inverting amplifier

Find the output voltage v_o, in terms of the input voltage v_1 and the external resistors.

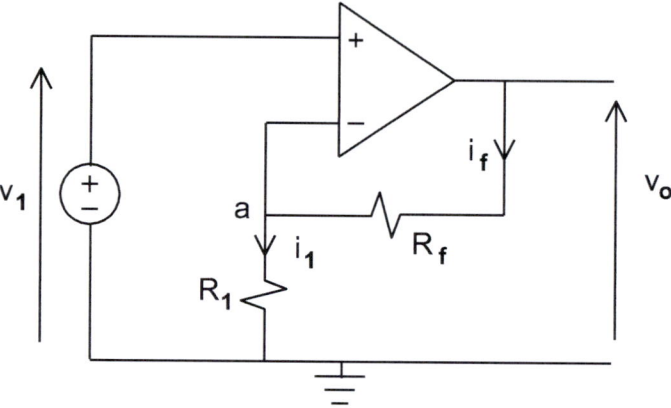

Fig. 8.7 Circuit for Example 8.2.

Since no current flows into the inverting terminal, by KCL at node a, the feedback current i_f equals the current through R_1.

$$i_f = i_1$$

The input terminals of the op amp are at the same potential. Since the non-inverting input is at v_1, then the potential of node a with respect to ground is also v_1. Ohm's law then gives

$$\frac{v_0 - v_1}{R_f} = \frac{v_1 - 0}{R_1}$$

$$v_0 = \left(1 + \frac{R_f}{R_1}\right) v_1$$

Example 8.3: Summing amplifier

Find the output voltage v_o, in terms of the input voltages v_1 and v_2, and the external resistors.

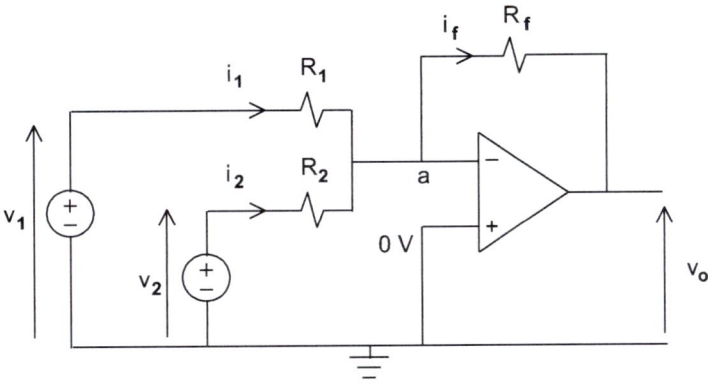

Fig. 8.8 Circuit for Example 8.3.

The terminals of the op amp are at the same potential and no current flows into the non-inverting terminal. Therefore, node a is at ground potential. KCL at node a, and Ohm's law give

$$i_1 + i_2 = i_f$$

$$\frac{v_1 - 0}{R_1} + \frac{v_2 - 0}{R_2} = \frac{0 - v_0}{R_f}$$

$$v_0 = -\left(\frac{R_f v_1}{R_1} + \frac{R_f v_2}{R_2} \right)$$

If $R_1 = R_2 = R_f$, then the output voltage equals the sum of the input voltages

$$v_0 = -(v_1 + v_2)$$

Example 8.4: Differential (subtracting) amplifier

Show that the output voltage is the difference of the input voltages.

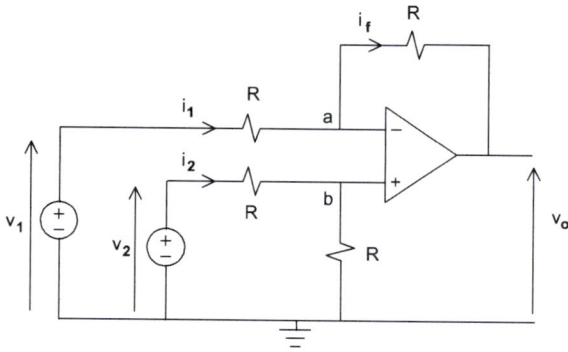

Fig. 8.9 Circuit for Example 8.4.

Nodes a and b are at the same potential. Let this potential be v with respect to ground. Also no currents flow into the terminals of the op amp. KCL at node a and Ohm's law give

$$\frac{v_1 - v}{R} = \frac{v - v_0}{R}$$
$$v = (v_1 + v_0)/2$$

Similarly, KCL at node b give gives a second expression for v

$$\frac{v_2 - v}{R} = \frac{v}{R}$$
$$v = v_2/2$$

Eliminating v gives

$$v_0 = v_2 - v_1$$

Example 8.5: Dividing amplifier

Show that the circuit of Fig. 8.10 acts as a divider so that $v_o = -v_1 / v_2$. The circuit uses a multiplier. Integrated circuit chips are available that perform multiplication, where the output voltage of the chip is proportional to the product of its inputs. The output voltage of the circuit v_o, and the input to the circuit v_2, are the inputs of the multiplier shown here. The multiplier is assumed to have a constant of proportionality of one.

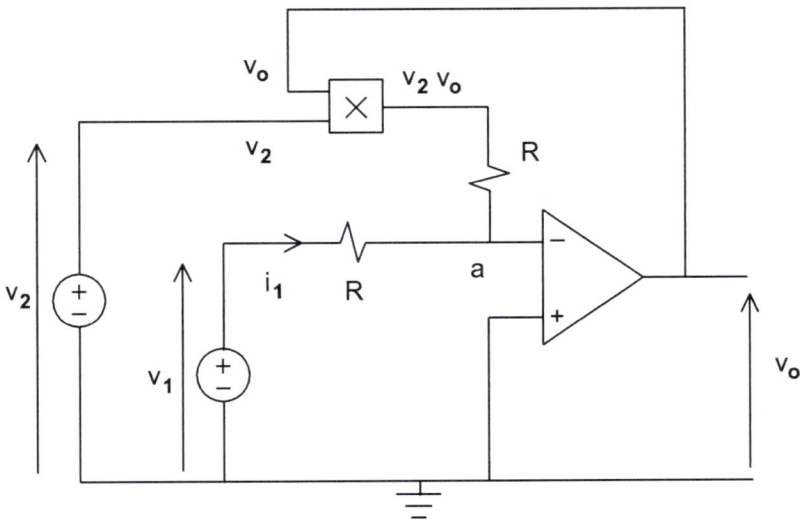

Fig. 8.10 Circuit for Example 8.5.

Node a is at zero potential and no current flows into the inverting input of the op amp. KCL at a and Ohm's law give

$$\frac{v_1 - 0}{R} = \frac{0 - v_2 v_0}{R}$$

$$v_0 = -\frac{v_1}{v_2}$$

Example 8.6: Active lowpass filter

The filters considered in Chapter 7 were not capable of amplifying the amplitudes of frequencies they allowed to pass to the output; they were passive filters. Show that the op amp circuit of Fig. 8.11 acts as an active lowpass filter that is capable of amplification in the passband.

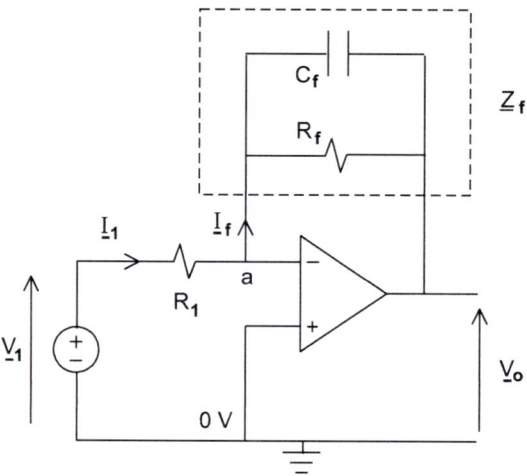

Fig. 8.11 Circuit for Example 8.6.

The ac characteristics of the circuit are of interest. The input voltage and the resulting quantities are represented as phasors. The impedance $\underline{Z}_f$, in the feedback loop, is the equivalent of the parallel combination of the capacitor, of capacitance C_f farads (and of impedance $\underline{Z}_c = 1/j\omega C_f$), and the resistor of resistance R_f. $\underline{Z}_f$ is then given by

$$\underline{Z}_f = \frac{1}{(1/\underline{Z}_c)+(1/R_f)} = \frac{R_f \underline{Z}_c}{R_f + \underline{Z}_c} = \frac{R_f / j\omega C_f}{R_f + 1/j\omega C_f} = \frac{R_f / \omega R_f C_f}{j + 1/\omega R_f C_f}$$

198

$$\underline{Z}_f = \frac{R_f / R_f C_f}{j\omega + 1/R_f C_f} = \frac{R_f \omega_h}{\omega_h + j\omega}$$

$\omega_h = 2\pi f_h = 1/R_f C_f$ = breakpoint frequency of the filter

(ω_h = high frequency cutoff frequency in radians/second

f_h = high frequency cutoff frequency in cycles/second).

As for the inverting amplifier of Example 8.1

$$\frac{V_0}{V_1} = -\frac{Z_f}{R_1}$$

Substituting for $\underline{Z}_f$ gives

$$\frac{V_0}{V_1} = -\frac{R_f \omega_h}{R_1 (\omega_h + j\omega)} = -\left(\frac{R_f}{R_1}\right)\left(\frac{1}{1 + jf / f_h}\right)$$

$$\left|\frac{V_0}{V_1}\right| = \frac{|V_0|}{|V_1|} = \left(\frac{R_f}{R_1}\right)\left(\frac{1}{\sqrt{1 + (f / f_h)^2}}\right)$$

The last equation is the magnitude of the transfer function of the circuit. It shows that for $f < f_h$, the term $(f / f_h)^2$ is small, compared to one, and can be ignored. For low frequencies the amplification then becomes R_f / R_1. By choosing $R_f > R_1$ sinusoids in the input signal with frequencies less than f_h pass to the output, with a gain of R_f / R_1. For $f > f_h$, $(f / f_h)^2 > 1$ and $V_0 < V_1$. Therefore, sinusoids in the input signal with frequencies greater than f_h are attenuated. The circuit is an active lowpass filter.

Example 8.7: Instrumentation amplifier

The circuit of Fig. 8.12 shows the basic concept behind an instrumentation-grade amplifier. It has a high input impedance. The output V_o depends on the difference of input voltages V_1 and V_2 i.e. the differential input signal. If the same voltage is applied to both inputs (the common-mode voltage), no voltage appears at the output. If the common-mode signal is noise, then the amplifier can detect small differential signals obscured by the noise. xR is a variable resistor that is used to vary the gain of the amplifier by changing the variable x. Moreover, whereas the input to the differential amplifier at the second stage of the circuit, V_{ab} is a floating input (not with respect to ground), the output voltage V_o is with respect to ground. Find the expression of the output voltage in terms of V_1, V_2 and x. Assume that $V_1 > V_2$.

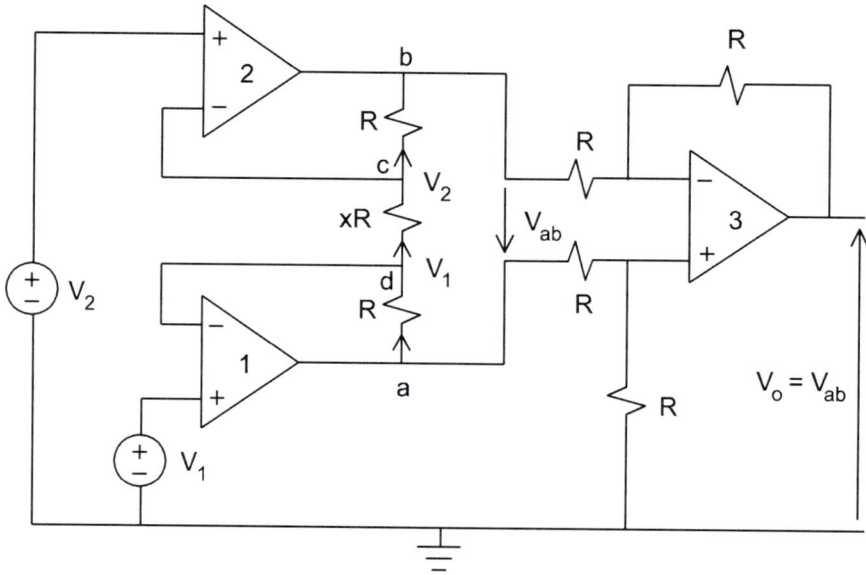

Fig. 8.12 Circuit of instrumentation amplifier for Example 8.7.

Both input terminals of op amp *1* are at V_1 and both inputs to op amp *2* at V_2. Therefore, node *d* is at V_1 and node *c* at V_2. Then the voltage difference appearing across the resistor of resistance xR is $V_1 - V_2$. The current through xR is $I = (V_1 - V_2)/xR$. Since no current enters the input terminals of op amps *1* and *2*, the current that flows through xR also flows through branches *cb* and *ad* in the directions shown in Fig. 8.12. Therefore, V_{ab} is given as

$$V_{ab} = I(2R + xR)$$
$$V_{ab} = \frac{(V_1 - V_2)}{xR}(2R + xR)$$
$$V_{ab} = [1 + (2/x)](V_1 - V_2)$$

V_{ab} is then the input to the differential amplifier *3* and, as in Example 8.4, the output voltage V_o is

$$V_o = V_{ab} = [1 + (2/x)](V_1 - V_2)$$

Chapter 9
Discrete Fourier Transform

9.1 Introduction

The discrete Fourier transform (DFT) is one of the most important signal processing algorithms used for the analysis of electrical signals. A continuous (analog) signal in the time domain gives the amplitude of the signal at any time. When this signal is input to the data acquisition (DAQ) board of the computer, the output from the DAQ board is a discrete sequence, or array, of samples. It is still a time domain representation of the signal but which now gives the amplitude of the original signal at the instants of time when it was sampled. The signal is no longer continuous, but made up of separate (discrete) data points. The DFT operates on the individual elements of the discrete array to transform the sequence from the time to the frequency domain. The frequency domain representation gives the spectrum of the signal. By computing the magnitude and phase of each frequency component present in the signal, the DFT gives the frequency or harmonic content of the original signal.

9.2 Sampled signal and the DFT

Assume a signal has been sampled at a rate of f_s samples per second. Then f_s is the sampling rate or sampling frequency. If the sampling interval, or time between adjacent samples, is δt seconds, then

$$\delta t = \frac{1}{f_s} \qquad (9.1)$$

If N samples have been obtained, where samples give the amplitude of the signal at spacings of δt seconds, then this is the time domain representation of the signal as a sequence, $x[i]$ where $i = 0, 1, 2......N\text{-}1$. The square brackets indicate that $x[i]$ is an array of discrete data points, or elements, where i is the zero based discrete time index. The DFT is then given by

$$X[m] = \sum_{i=0}^{i=N-1} x[i] e^{-j2\pi i m/N} \qquad m\text{=}0,1,2....N\text{-}1 \ \text{ and } \ j = \sqrt{-1} \qquad (9.2)$$

The DFT operates on the N samples of $x[i]$ to transform it into an array, $X[m]$, in the frequency domain. $X[m]$ is composed of N samples which give the frequency components present in the signal. m is the discrete frequency index corresponding to each component $X[m]$ of the DFT. The spacing between adjacent frequencies, or frequency resolution, δf Hz, in the frequency domain representation of the signal, is

$$\delta f = \frac{1}{N\delta t} = \frac{f_s}{N} \qquad (9.3)$$

Fig. 9.1(i) shows a time domain representation of one cycle of a sine wave with $N = 30$ samples. Fig. 9.1(ii) shows the 30 samples

203

of the frequency domain representation of the sine wave of Fig. 9.1(i) after the DFT has been taken using Equ. 9.2. Two frequency components appear in Fig. 9.1(ii), at $m = 1$ and at $m = 29$. Both components have a magnitude of 90. The spectrum of Fig. 9.1(ii) contains duplicate information; it is symmetrical about $m=N/2=15$.

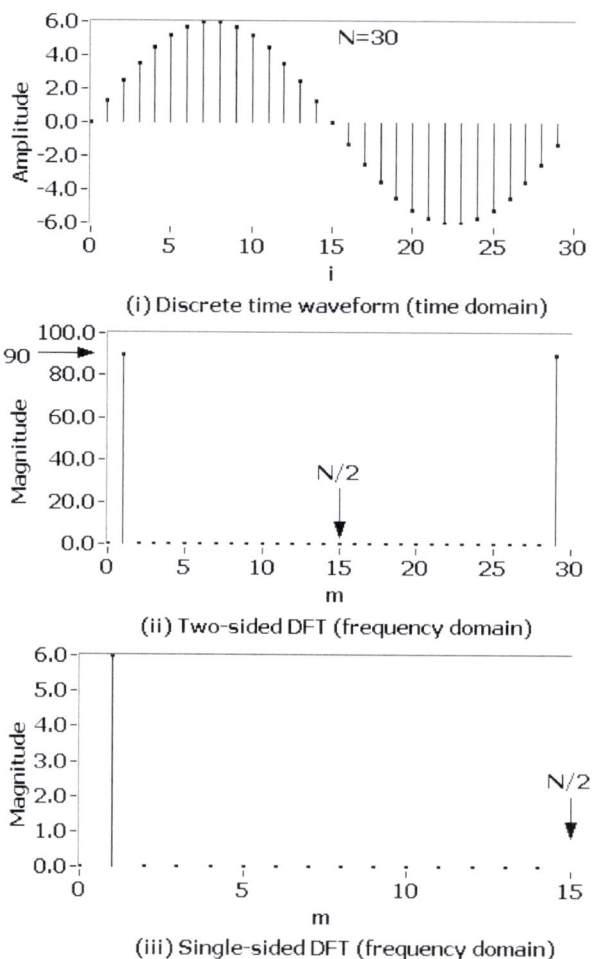

(i) Discrete time waveform (time domain)

(ii) Two-sided DFT (frequency domain)

(iii) Single-sided DFT (frequency domain)

Fig. 9.1 Time domain representation in (i) and frequency domain representation in (ii) and (iii) for a sampled sine wave. (ii) Two sided DFT of sine wave. (iii) Single sided DFT of sine wave.

Frequency components above $m = N/2 = 15$ in the two sided DFT of Fig. 9.1(ii) are considered negative frequencies and are discarded.

Ignoring the frequency components above $N/2$ results in the one sided DFT, shown in Fig. 9.1(iii), of the sine wave of Fig. 9.1(i). To accurately compute the magnitude of frequency components in the one sided DFT spectrum, the magnitude of each component below $N/2$, excluding the dc component, is doubled and then divided by N. In the case of Fig. 9.1(ii), there is one frequency component, at $m = 1$, which is above dc and below $m = N/2 = 15$, and it has a magnitude of 90. Therefore, the magnitude of the frequency component in the one sided DFT is $(2 \times 90)/30 = 6$. This restores the magnitude of the frequency component to that expected from the signal seen in the time domain. If a dc component is present in the spectrum (given by the $X(0)$ term), it need only be divided by N to restore its magnitude to its true value.

Therefore, using the DFT of Equ. 9.2, the N samples of the input signal, $x[i]$ are transformed from the time domain into N components $X[m]$ in the frequency domain. Except for the dc term $X[0]$, the $X[m]$ terms are complex; each has a real and an imaginary part or a magnitude and a phase angle. If $x[i]$ is real, then $X[m]$ is symmetrical about the index $N/2$. The index $N/2$ is the Nyquist index.

9.3 Normalized frequency and frequency resolution

In digital signal processing frequencies are considered in reference to the sampling frequency. This results in the normalized frequency. If an analog signal has a frequency of f_a Hz, then its normalized frequency, after it is sampled at a rate of f_s samples per second, is f_n where

$$f_n = \frac{f_a}{f_s} \qquad (9.5)$$

f_n has units of (cycles/second)/(samples/second)=cycles/sample.

Figure 9.2(i) shows the discrete time sequence of a sine wave, with an analog frequency of 100 cycles per second, which was sampled at a rate of 1000 samples per second. Figure 9.2(ii) is the magnitude plot of the one sided DFT of the sine wave of Fig. 9.2(i) where the magnitude of the frequency component has been doubled and divided by N to give it its correct value. The sine wave had an amplitude before sampling of unity.

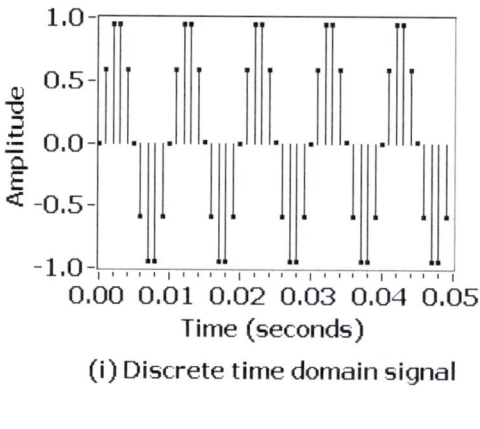

(i) Discrete time domain signal

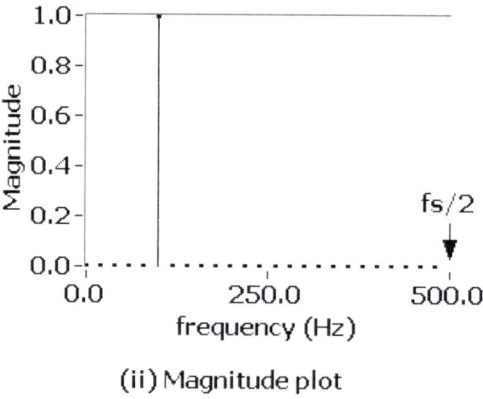

(ii) Magnitude plot

Fig. 9.2 (i) Time domain sequence of a sampled sine wave. (ii) Frequency domain sequence of single sided DFT of sine wave in (i). Time axis in (i) and frequency axis in (ii) have been calibrated.

The normalized frequency is $f_n = 100/1000 = 0.1$ cycles per sample. $N = 50$ samples were taken resulting in $50 \times 0.1 = 5$ cycles, as displayed in Fig. 9.2(i). The time spacing $\delta t = 1/f_s$ is 0.001 seconds. Therefore, the total time taken to scan for the 50 samples was $\delta t \times N = 0.05$ seconds.

The frequency resolution for the magnitude spectrum of Fig. 9.2(ii) is $\delta f = f_s / N = 1000/50 = 20$ Hz. With a sampling rate of 1000 samples per second the DFT can detect frequencies in the input signal which are between dc and just under $f_s / 2 = 500$ Hz, with a resolution between frequencies of 20 Hz. The magnitude of unity for the frequency component at 100 Hz in Fig. 9.2(ii) corresponds to the $X(5)$ term of the DFT and gives the magnitude of the 100 Hz component present in the input signal. For this input signal, the magnitudes for frequencies corresponding to other $X(m)$ terms are, as expected, zero.

The frequency corresponding to $f_s / 2$ is the Nyquist frequency. (In the above example, this corresponds to the frequency index $m = N/2 = 25$ in Fig. 9.2(ii)). The Nyquist frequency is linked to the sampling theorem. According to the sampling theorem, to detect a sinusoid with a frequency of f Hz in a signal, the signal needs to be sampled at a rate equal to or higher than $2f$ samples per second. If a signal is sampled at f_s samples per second, then the only meaningful frequencies that can be detected by the DFT are those just above dc and just below the Nyquist frequency of $f_s / 2$ Hz.

9.4 Power spectrum

Squaring the magnitude of each component in the magnitude plot of the DFT gives the power contained in each $X[m]$ term. The resulting plot is the power spectrum.

Example 9.1: DFT of signal with _N_ odd

Determine the DFT of the sequence of seven sample points shown in Fig. 9.3(i).

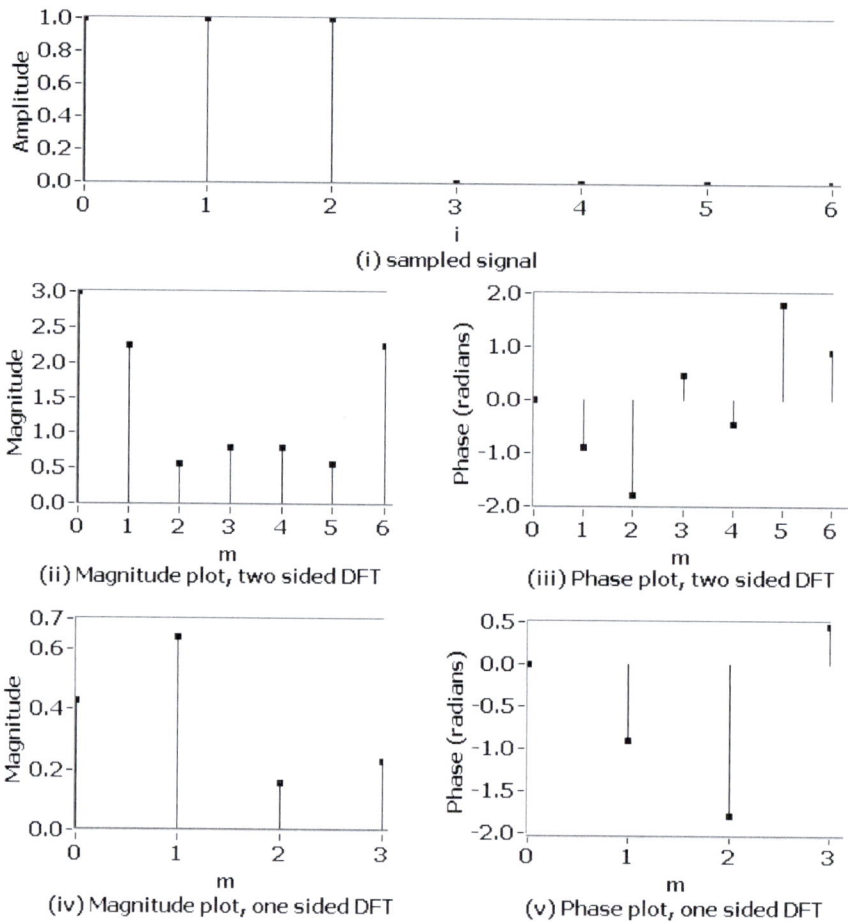

(i) sampled signal

(ii) Magnitude plot, two sided DFT

(iii) Phase plot, two sided DFT

(iv) Magnitude plot, one sided DFT

(v) Phase plot, one sided DFT

Fig. 9.3 Example 9.1.

Starting with Equ. 9.2

$$X[m] = \sum_{i=0}^{i=N-1} x[i]e^{-j2\pi im/N} \qquad \text{where } m=0,1,2.....6$$

$$X[0] = 1 \times e^{-j\frac{2\pi}{7} \times 0 \times 0} + 1 \times e^{-j\frac{2\pi}{7} \times 1 \times 0} + 1 \times e^{-j\frac{2\pi}{7} \times 2 \times 0} + 0 + 0 + 0 + 0$$

$$X[0] = 3$$

$$X[1] = 1 \times e^{-j\frac{2\pi}{7} \times 0 \times 1} + 1 \times e^{-j\frac{2\pi}{7} \times 1 \times 1} + 1 \times e^{-j\frac{2\pi}{7} \times 2 \times 1} + 0 + 0 + 0 + 0$$

$$X[1] = 2.247e^{-j0.898}$$

$$X[2] = 1 \times e^{-j\frac{2\pi}{7} \times 0 \times 2} + 1 \times e^{-j\frac{2\pi}{7} \times 1 \times 2} + 1 \times e^{-j\frac{2\pi}{7} \times 2 \times 2} + 0 + 0 + 0 + 0$$

$$X[2] = 0.555e^{-j1.795}$$

$$X[3] = 1 \times e^{-j\frac{2\pi}{7} \times 0 \times 3} + 1 \times e^{-j\frac{2\pi}{7} \times 1 \times 3} + 1 \times e^{-j\frac{2\pi}{7} \times 2 \times 3} + 0 + 0 + 0 + 0$$

$$X[3] = 0.802e^{j0.449}$$

$$X[4] = 1 \times e^{-j\frac{2\pi}{7} \times 0 \times 4} + 1 \times e^{-j\frac{2\pi}{7} \times 1 \times 4} + 1 \times e^{-j\frac{2\pi}{7} \times 2 \times 4} + 0 + 0 + 0 + 0$$

$$X[4] = 0.802e^{-j0.449}$$

$$X[5] = 1 \times e^{-j\frac{2\pi}{7} \times 0 \times 5} + 1 \times e^{-j\frac{2\pi}{7} \times 1 \times 5} + 1 \times e^{-j\frac{2\pi}{7} \times 2 \times 5} + 0 + 0 + 0 + 0$$

$$X[5] = 0.555e^{j1.795}$$

$$X[6] = 1 \times e^{-j\frac{2\pi}{7} \times 0 \times 6} + 1 \times e^{-j\frac{2\pi}{7} \times 1 \times 6} + 1 \times e^{-j\frac{2\pi}{7} \times 2 \times 6} + 0 + 0 + 0 + 0$$

$$X[6] = 2.247e^{j0.898}$$

$X[0]$ is the dc term. $X[1]$ is the complex conjugate of $X(6)$ i.e. $X[1] = X^*[6]$. Similarly, $X[2] = X^*[5]$, $X[3] = X^*[4]$. Therefore, $X[4], X[5], X[6]$ could have been written after $X[1], X[2], X[3]$

had been calculated. Figures 9.3(ii) and (iii) show the magnitude and phase plots, respectively, of the two sided DFT that was calculated. The magnitude plot shows even symmetry while the phase plot is odd symmetric. Figures 9.3(iv) and (v) show the magnitude and phase plots of the one sided DFT with the magnitudes adjusted. Since the number of samples ($N=7$) is odd, there is no integer $m=N/2$ corresponding to the Nyquist index. Therefore, there is no $X[m]$ term at the Nyquist frequency.

Example 9.2: DFT of signal with *N* even

Determine the DFT of the sequence of eight sample points shown in Fig. 9.4(i).

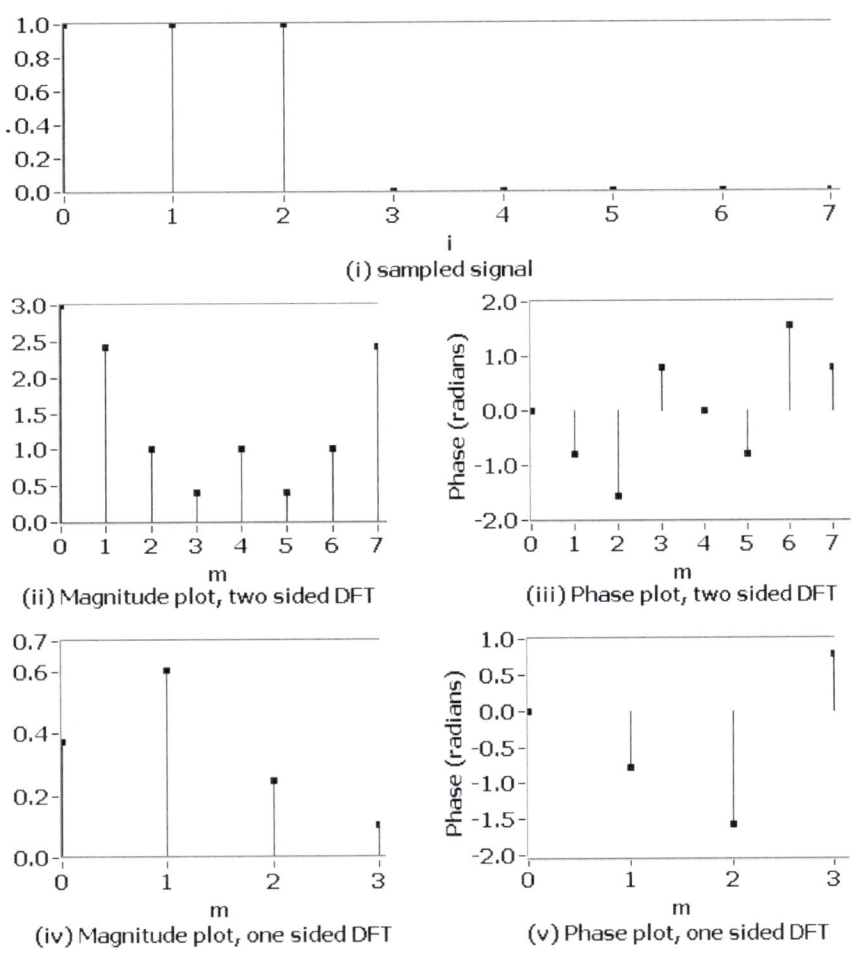

(i) sampled signal

(ii) Magnitude plot, two sided DFT

(iii) Phase plot, two sided DFT

(iv) Magnitude plot, one sided DFT

(v) Phase plot, one sided DFT

Fig. 9.4 Example 9.2.

Starting with Equ. 9.2

$$X[m] = \sum_{i=0}^{i=N-1} x[i]e^{-j2\pi im/N} \qquad \text{where } m=0,1,2.....7$$

$X[0] = 1 \times e^{-j\frac{2\pi}{8} \times 0 \times 0} + 1 \times e^{-j\frac{2\pi}{8} \times 1 \times 0} + 1 \times e^{-j\frac{2\pi}{8} \times 2 \times 0} + 0 + 0 + 0 + 0 + 0$

$X[0] = 3$

$X[1] = 1 \times e^{-j\frac{2\pi}{8} \times 0 \times 1} + 1 \times e^{-j\frac{2\pi}{8} \times 1 \times 1} + 1 \times e^{-j\frac{2\pi}{8} \times 2 \times 1} + 0 + 0 + 0 + 0 + 0$

$X[1] = 2.414e^{-j\pi/4} = 2.414e^{-j0.785}$

$X[2] = 1 \times e^{-j\frac{2\pi}{8} \times 0 \times 2} + 1 \times e^{-j\frac{2\pi}{8} \times 1 \times 2} + 1 \times e^{-j\frac{2\pi}{8} \times 2 \times 2} + 0 + 0 + 0 + 0 + 0$

$X[2] = 1e^{-j\pi/2} = 1e^{-j1.571}$

$X[3] = 1 \times e^{-j\frac{2\pi}{8} \times 0 \times 3} + 1 \times e^{-j\frac{2\pi}{8} \times 1 \times 3} + 1 \times e^{-j\frac{2\pi}{8} \times 2 \times 3} + 0 + 0 + 0 + 0 + 0$

$X[3] = 0.414e^{j\pi/4} = 0.414e^{j0.785}$

$X[4] = 1 \times e^{-j\frac{2\pi}{8} \times 0 \times 4} + 1 \times e^{-j\frac{2\pi}{8} \times 1 \times 4} + 1 \times e^{-j\frac{2\pi}{8} \times 2 \times 4} + 0 + 0 + 0 + 0 + 0$

$X[4] = 1e^{-j0}$

$X[5] = 1 \times e^{-j\frac{2\pi}{8} \times 0 \times 5} + 1 \times e^{-j\frac{2\pi}{8} \times 1 \times 5} + 1 \times e^{-j\frac{2\pi}{8} \times 2 \times 5} + 0 + 0 + 0 + 0 + 0$

$X[5] = 0.414e^{-j\pi/4} = 0.414e^{-j0.785}$

$X[6] = 1 \times e^{-j\frac{2\pi}{8} \times 0 \times 6} + 1 \times e^{-j\frac{2\pi}{8} \times 1 \times 6} + 1 \times e^{-j\frac{2\pi}{8} \times 2 \times 6} + 0 + 0 + 0 + 0 + 0$

$X[6] = 1e^{j\pi/2} = 1e^{j1.571}$

$X[7] = 1 \times e^{-j\frac{2\pi}{8} \times 0 \times 7} + 1 \times e^{-j\frac{2\pi}{8} \times 1 \times 7} + 1 \times e^{-j\frac{2\pi}{8} \times 2 \times 7} + 0 + 0 + 0 + 0 + 0$

$X[7] = 2.414e^{j\pi/4} = 1e^{j0.785}$

Figures 9.4(ii) and (iii) show the magnitude and phase plots, respectively, of the two sided DFT that was calculated. Figures

212

9.4(iv) and (v) show the magnitude and phase plots of the one sided DFT with the magnitudes adjusted. Since the number of samples ($N=8$) is even, the integer $m=N/2=4$ corresponds to the Nyquist index and the $X[4]$ term is at the Nyquist frequency.

Chapter 10
The diode

10.1 Introduction

The diode is a device that finds application in many useful circuits. It is a two terminal device with an anode and a cathode. If the anode is more positive than the cathode, a current passes through the diode. If a voltage is applied across the diode so that the anode is negative with respect to the cathode, then no current flows through the diode and the device acts as an open circuit. This property of either being on (conducting) or off (not conducting) means that the diode can be used in applications such as rectifying ac to dc, switching, voltage doubling, waveform shaping and demodulating. The ideal diode is introduced and some diode circuits are described in this chapter.

10.2 Diode characteristics

The symbol of the diode is given in Fig. 10.1(ii). Conduction is due to electrons and holes in the pn junction semiconductor diode. The p type material is the anode and the n side of the junction is the cathode. The cathode is usually marked with a band on the actual diode.

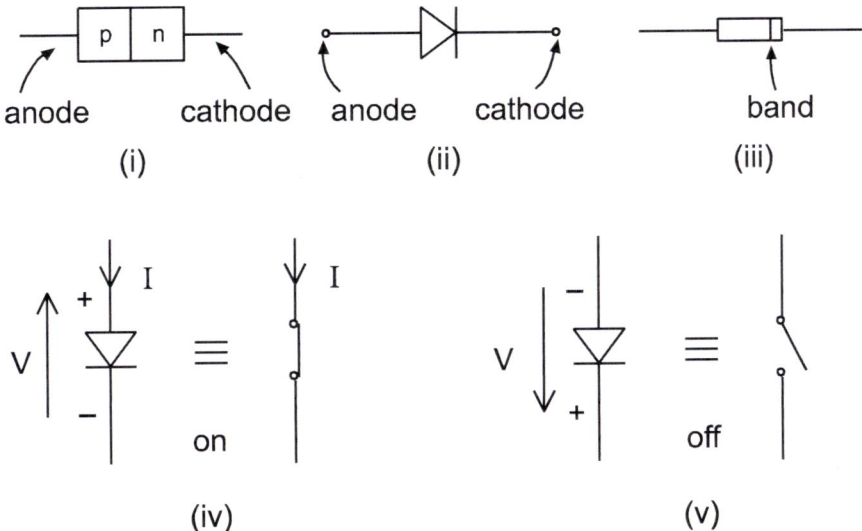

Fig. 10.1 (i) pn junction diode (ii) diode symbol (iii) band identifies cathode (iv) forward biased ideal diode (on state, diode conducting current) (v) reverse biased diode (off state, represented by an open switch).

When a voltage difference is applied across the diode so that the anode is more positive than the cathode a current flows through the device as shown in Fig. 10.1(iv). The diode is then in forward bias, and is said to be on. When the anode is negative with respect to the cathode, as in Fig. 10.1(v), the diode is in reverse bias or off. It is then represented as an open switch.

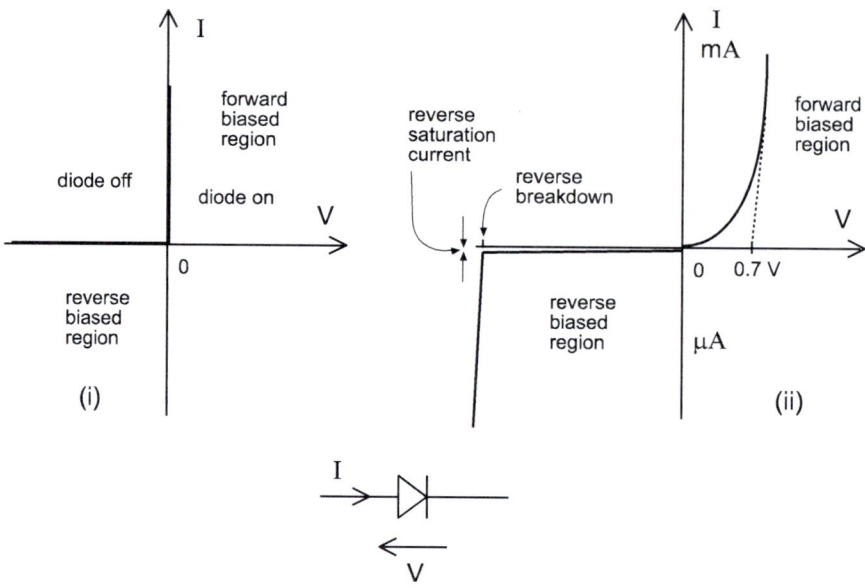

Fig. 10.2 Current-voltage characteristics of (i) ideal (ii) real diode.

The current-voltage (I-V) characteristics of a practical silicon diode appear in Fig. 10.2 (ii). The forward bias voltage must exceed 0.7 V, at room temperature, before the forward current starts to increase rapidly with increase in forward voltage. A voltage of about 0.7 volts then appears across the conducting diode. In reverse bias, a small leakage current, the reverse saturation current, flows through the diode. As the reverse voltage is increased the diode undergoes reverse breakdown with a rapid increase in current. The scales for the positive and negative voltages in Fig. 10.2(ii) are not the same, being of the order of a volt in forward bias and tens of volts in reverse bias.

Fig. 10.2(i) shows the I-V characteristics of the ideal diode. Here the diode conducts as long as a forward bias (of whatever magnitude) is applied across it. The ideal diode in forward bias is represented as a short circuit (Fig. 10.1(iv)). In reverse bias the ideal diode is an open switch (Fig. 10.1(v)).

216

10.3 Half-wave rectifier

A rectifier converts ac to dc. The ac source in the circuit of Fig. 10.3(i) supplies the sinusoidal voltage shown in Fig. 10.3(ii). In the positive half-cycle, the diode is forward biased and conducts. Assuming an ideal diode, the supply voltage then appears across the resistor as the output voltage $v_R(t)$. During the negative half-cycle the diode is in reverse bias and acts as an open circuit. The output voltage $v_R(t)$ is then zero. The output voltage, as a function of time, is shown in Fig. 10.3(iii). The circuit acts as a half-wave rectifier.

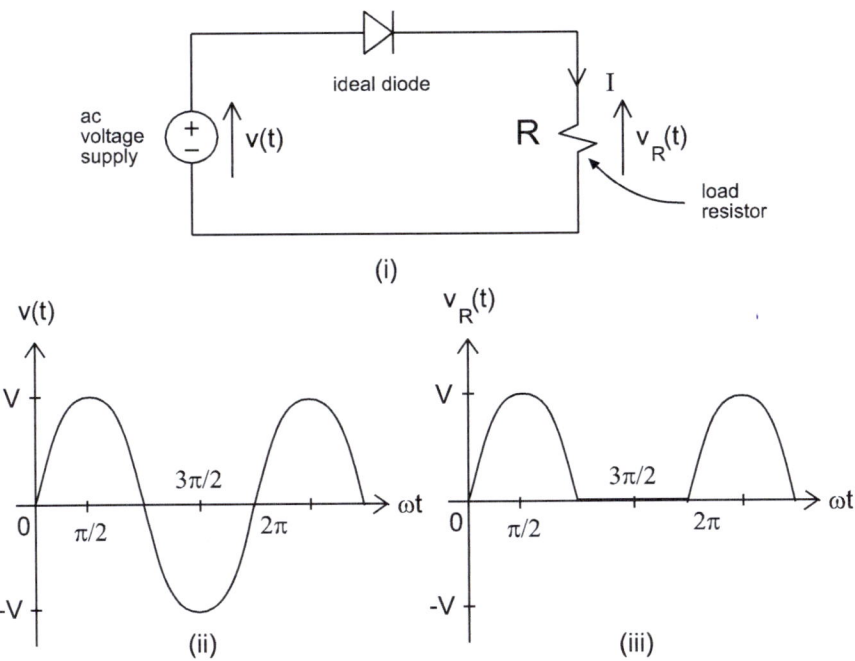

Fig. 10.3 Half-wave rectifier circuit.

10.4 Half-wave rectifier with smoothing capacitor

The inclusion of a smoothing capacitor in the half-wave rectifier circuit results in the circuit of Fig. 10.4(i). This circuit converts ac to a nearly constant dc waveform. The ac input voltage can be supplied by the secondary of a transformer. In the first half of the positive half-cycle of the input voltage $v(t)$, the diode is forward biased. Assuming an ideal diode the supply voltage appears across the load resistor so that $v_R(t) = v(t)$. Current also flows through the capacitor and charges it to a peak voltage of V, the amplitude of the

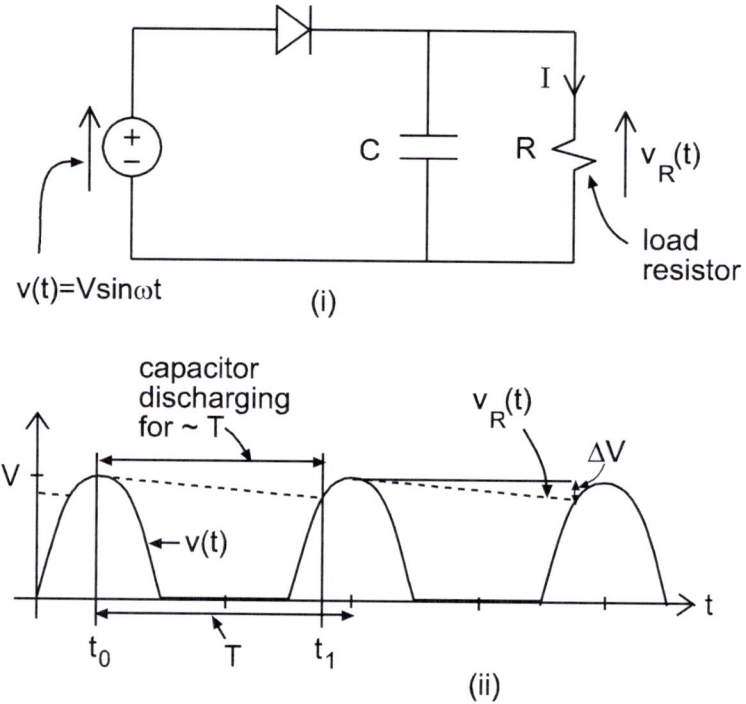

v(t)=Vsinωt   (i)

(ii)

Fig. 10.4 Converting ac voltage to nearly dc.

supply voltage. After reaching the amplitude V, at time t_0 in Fig. 10.4(ii), $v(t)$ falls below V. Since the voltage across the capacitor is

218

still V, the diode is now in reverse bias and represented by an open circuit. No current flows through the diode but the capacitor starts discharging through the load resistor. The resulting current through the resistor ensures that the voltage across the resistor (the output voltage) does not fall to zero. This voltage $v_R(t)$ is shown as the dashed curve in Fig. 10.4(ii). The capacitor discharges for almost the duration of a period T of the supply voltage. The diode is in forward bias once again at t_1 when the supply voltage rises above the capacitor voltage and the capacitor starts charging again. This charging stops, once again, when the capacitor voltage reaches V. During the discharging phase the capacitor loses charge and the voltage across it and, therefore, across the load resistor decreases by a total of ΔV (Fig. 10.4(ii)). This change in the load voltage is the ripple. If the ripple is kept small, the voltage across the load resistor can be used as a good approximation of a constant dc voltage.

Assuming that the capacitor discharges for a time T, when it supplies an average current I through the resistor, then the charge supplied during this time is ΔQ, where

$$\Delta Q = IT = C\Delta V \tag{10.1}$$
$$C = IT / \Delta V \tag{10.2}$$

For $V = 58$ V, $I = 0.05$ A, $T = 0.017$ seconds ($f = 60$Hz) and a ripple of 1%, Equ. 10.2 gives

$$C = \frac{0.05 \times (1/60)}{0.01 \times 58} = 1437 \ \mu F$$

Changing the ripple to 5%, in this example, gives C as 287 μF. These are large value capacitors capable of storing large amounts of charge. The release of this charge is accompanied with large currents that can deliver very nasty shocks. Also, inductors are capable of creating large and dangerous voltages. Great care should be exercised when using these in the laboratory.

10.5 A diode application in the astable multivibrator

The monostable and then the astable multivibrator will first be described followed by the use of the diode with the latter in improving the range of its mark-to-space ratio.

10.5.1 Monostable multivibrator

A monostable multivibrator outputs a pulse, of width T, when the circuit is triggered by a narrow input pulse. A monostable has one stable state characterized by an output voltage $V_0 =$ low (LO). This is a voltage close to zero. When the trigger pulse is applied, V_0 goes high (HI). This is a voltage close to, but below, the supply voltage in the present case. The monostable is in its unstable state for the duration of the pulse. The 555 timer is used here to describe monostable action. The resistor and capacitor external to the 555 IC decide the pulse width T.

The main features of the 555 timer IC are shown in the dashed rectangle of Fig. 10.5. V_{cc} is the external dc supply voltage. Three internal resistors, each $R = 5k\Omega$ (5000Ω), form a voltage divider so that $(2/3)V_{cc}$ volts appear at the inverting input of comparator 1 and $(1/3)V_{cc}$ at the non-inverting input of comparator 2. The trigger input is connected to the inverting input of comparator 2. With no trigger pulse applied, the trigger input is at V_{cc} so that the inverting input is at a higher potential than the noninverting input of comparator 2, the output of comparator 2 is LO, and the output of the flip flop $\overline{Q}$ is HI. This means that the base terminal b, of the discharge transistor is HI and the transistor is on with current flowing from the collector c to the emitter e and then to ground. This means that in this state, current bypasses the external capacitor C, and prevents it from charging and developing a voltage across it, as the current flows from the supply V_{cc} through the external resistor R_A and the transistor to ground. Pin 7 is close

220

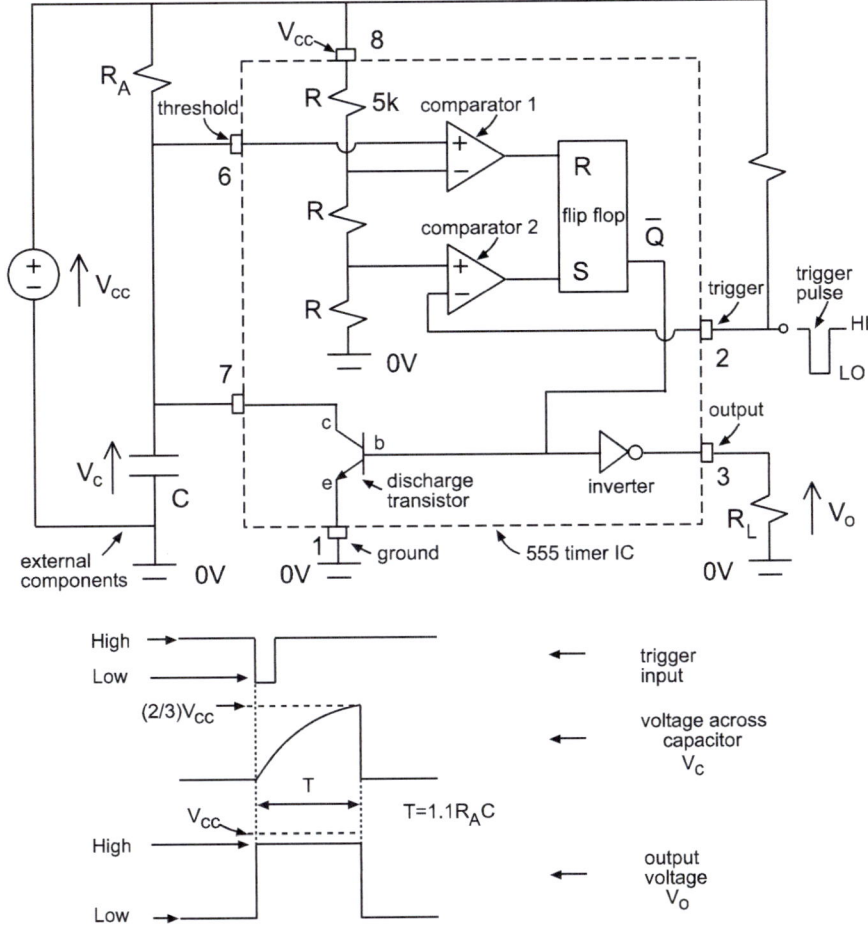

Fig. 10.5 Monostable multivibrator circuit using 555 timer.

to zero volts. $\overline{Q} = $ HI is also input to the inverter so that the output voltage V_0 = LO. This describes the monostable in its stable state.

When a negative going trigger pulse is applied to pin 2, so that the inverting input of comparator 2 becomes LO, the output of comparator 2 goes positive momentarily and this sets the flip flop so that its output $\overline{Q}$ goes low. The discharge transistor goes off and the output voltage goes HI. The transistor is no longer a conducting path to ground and the capacitor C starts charging, through R_A, towards V_{cc}. When V_c, the voltage across the capacitor, reaches the threshold of $(2/3)V_{cc}$, the output of comparator 1 goes to HI. This resets the flip flop so that $\overline{Q}$ is HI and the output voltage V_o is LO. The transistor starts to conduct and rapidly discharges the capacitor.

During charging, the capacitor charges from 0 to $(2/3)V_{cc}$, for T seconds, while tending to a final steady state voltage of V_{cc}. Then, from Equ. 6.15

$$V_c(T) = (2/3)V_{cc} = [0 - V_{cc}]e^{-\frac{T}{R_A C}} + V_{cc} \tag{10.1}$$

$$T = 1.1 R_A C \tag{10.2}$$

10.5.2 Astable multivibrator

The output of the astable multivibrator fluctuates between two, nonstable, voltage levels. The astable multivibrator circuit is shown in Fig. 10.6. It generates a series of pulses. Pins 2 and 6 are connected together. This allows the capacitor to charge and discharge between the threshold and trigger levels of $(1/3)V_{cc}$ and $(2/3)V_{cc}$, respectively. C charges through R_A and R_B towards V_{cc}, when the transistor is off. The output is HI during this charging time of t_1 seconds. When the voltage across C reaches $(2/3)V_{cc}$, the transistor turns off and the output goes LO as C starts discharging through R_B towards zero volts. The output stays LO for the discharging time of t_2 seconds. When the voltage across C reaches $(1/3)V_{cc}$ the capacitor starts charging and the output goes HI again as a new cycle begins.

The resulting train of pulses has a period of T $(=t_1 + t_2)$ and a pulse repetition rate of $1/T$ pulses per second (pps). The duty cycle is $(t_1/T)100\%$ and (t_1/t_2) is the mark-to-space ratio.

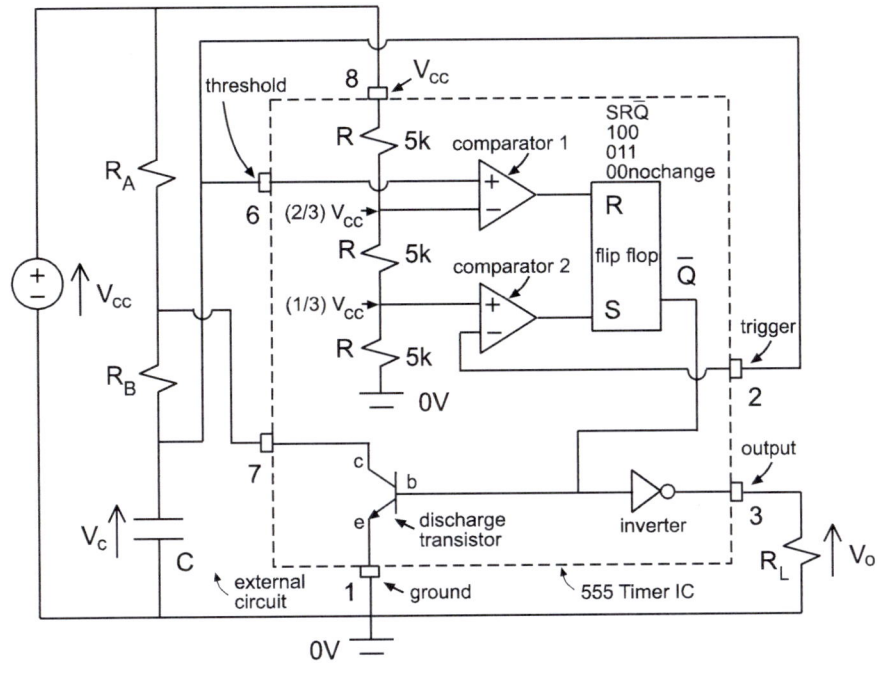

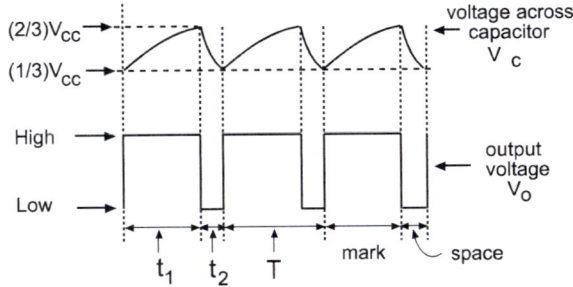

Fig. 10.6 Astable multivibrator circuit.

At the end of charging, Equ. 6.15 gives

$$V_c(t_1) = (2/3)V_{cc} = [(V_{cc}/3) - V_{cc}]e^{-\frac{t_1}{(R_A+R_B)C}} + V_{cc} \qquad (10.3)$$

$$t_1 = 0.69C(R_A + R_B) \qquad (10.4)$$

At the end of discharging, Equ. 6.15 gives

$$V_c(t_2) = (1/3)V_{cc} = [(2/3)V_{cc} - 0]e^{-\frac{t_2}{R_B C}} + 0 \tag{10.5}$$

$$t_2 = 0.69CR_B \tag{10.6}$$

The expressions for t_1 and t_2 show that $t_1 > t_2$ and, therefore, the duty cycle $(t_1/T)100 = [t_1/(t_1 + t_2)]100 > 50\%$. To increase the range of the mark-to-space ratio, and of the duty cycle, two diodes are used with the circuit of Fig. 10.6, as seen in Fig. 10.7.

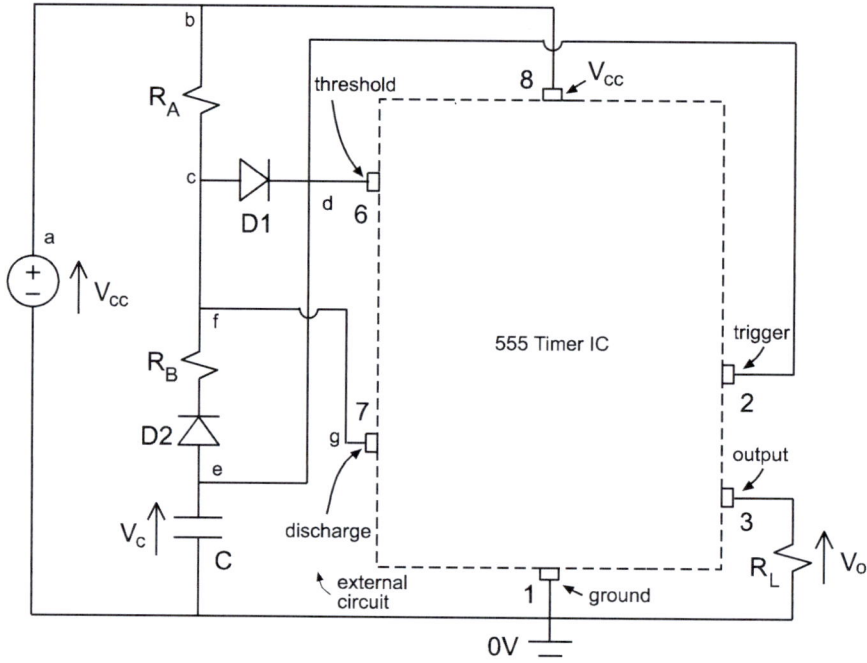

Fig. 10.7 Astable multivibrator circuit with control over duty cycle.

Diode D1 conducts while C is charging through path *abcde*, and D2 conducts when C is discharging through path *efg*. Here, $t_1 = 0.69CR_A$ and $t_2 = 0.69CR_B$, and duty cycles above or below 50% are possible.

Chapter 11
Elements of Energy Conversion Systems

11.1 Introduction

The application of electromagnetic induction in the electric generator and motor is discussed in this chapter together with the characteristics of the permanent magnet dc motor and three phase power. Electrical conduction in metals, semiconductors and photoconductors and the characteristics of solar cells and of batteries are included.

11.2 Electric generator and motor

A conductor moving in a magnetic field and cutting lines of magnetic force has, by Faraday's law of electromagnetic induction, a voltage generated across it. If the conductor is part of a loop in a circuit, then a current will flow in the loop. This is the basis of the electric generator. Conversely, when a current is passed by an external source through a stationary conductor in a magnetic field, then a force is generated that acts on the conductor; the consequent movement of the conductor is utilized in the electric motor.

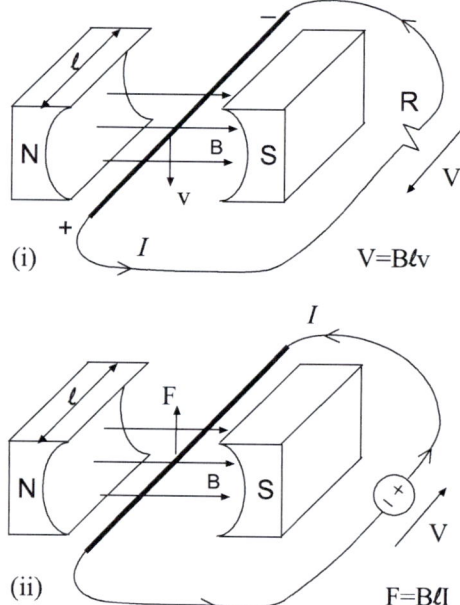

Fig. 11.1 Generation of (i) voltage and (ii) movement

The conductor in Fig. 11.1(i) moves with a uniform velovity of v meters/second (m/s) at right angles to the magnetic field of flux density of B webers/m^2 (or teslas) due to a permanent magnet. If the length of the conductor in the field is l, then a voltage is generated between the ends of the conductor that pushes a current through the circuit. The voltage generated is $V = Blv$. Note that the vectors representing $\overline{B}$, $\overline{v}$ and the axis of the conductor are mutually perpendicular. The polarity of V is such that if the conductor falls in the palm of the right hand, while the fingers point in the direction of the magnetic field, then the thumb points towards the positive terminal of the generator. The direction of the magnetic field is the direction that a unit north pole would move in if it were released between the pole pieces of the magnet.

The conductor in Fig. 11.1(ii) is in a magnetic field and has a current I passing through it due to the external voltage source V. It then experiences a force F. In the figure the current is perpendicular to the magnetic field and then $F=BlI$ where the magnetic field, the current and the force are mutually perpendicular. The direction of the force is found by imagining the current to enter the palm of the right hand while the fingers point in the direction of the magnetic field. The thumb then points in the direction of the force.

For the above principles to provide any kind of contiuous generator or motor action consider the electromagnetic devices of Fig.11.2(i) where a coil of conductor wire $abcd$ rotates at a constant angular velocity ω in a clockwise direction between the polepieces of a permanent magnet. Each end of the coil is connected to a slip ring. A stationary carbon brush presses against a ring to make electrical contact with it. In the position shown in the figure sides ab and cd are at right angle to the magnetic field and are cutting the maximum number of lines of force. This results in a maximum voltage V_{ad} being generated across the terminals of the generator. As the coil rotates a further 90° it is no longer cutting lines of force and the output voltage falls to zero. The magnitude of the output voltage between these two positions of the coil lies between the maximum voltage and zero. In the next 90° turn of the coil the polarity of the output voltage switches so that V_{ad} is now negative. The output voltage is a sinusoid for every 360° turn of the coil. Replacing slip rings with the half rings in the commutator of Fig. 11.1(ii) results in a pulsing output that is nontheless always positive. Using multiple coils each with its

227

segment in the commutator taps the positive voltage maxima of each coil resulting in a dc voltage at the output in the dc generator.

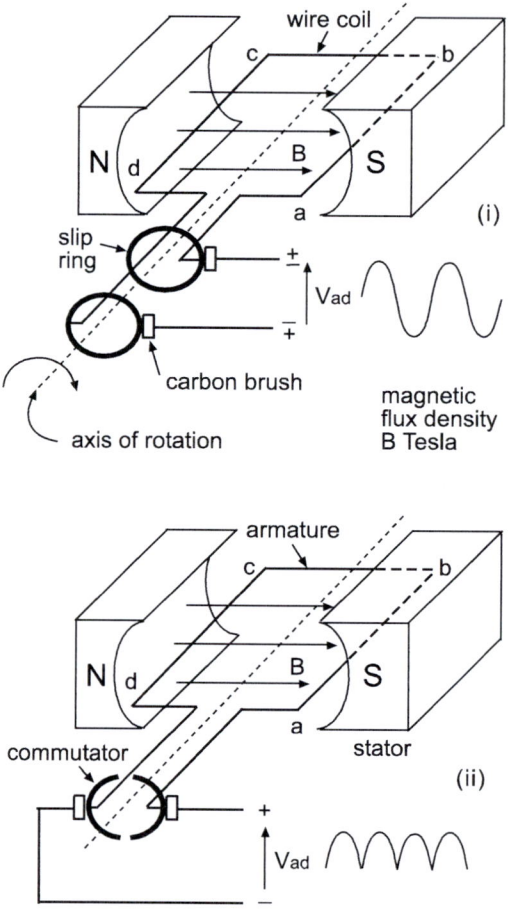

Fig. 11.2 Electromagnetic machine

Appling a dc voltage to the terminals of a dc generator results in the flow of current in the coil or armature. The interaction of the armature current with the magnetic field results in rotation of the armature and the machine now acts as a motor.

11.3 DC motor characteristics

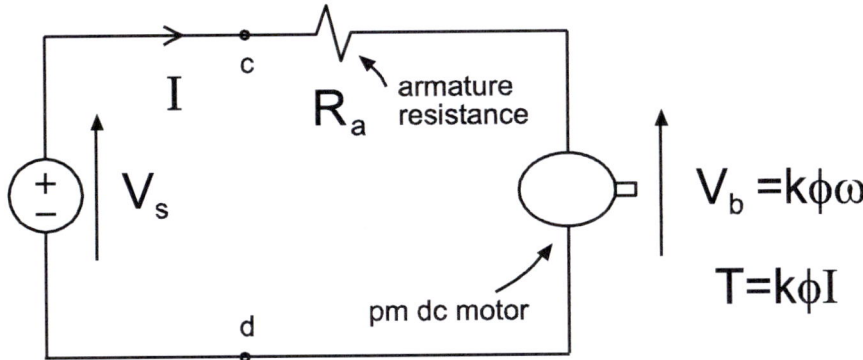

Fig. 11.3 Supply voltage applied across pm dc motor

The equivalent model of a permanent magnet (pm) dc motor is shown to the right of terminals c and d in Fig. 11.3. R_a is the armature resistance. The externally applied supply voltage V_s makes the armature (rotor) spin at an angular velocity of ω radians/second. The stationary part (stator) of the motor houses the permanent magnet that provides the magnetic field. The motor can then do mechanical work through the torque T (in Newton-meters, Nm) that is developed. The torque depends on the area A (radius r times length l) of the armature, the number of turns n in the armature coil, the flux density B and of the armature current I. The torque is given by

$$T = 2nrBlI = nABI \qquad (11.1)$$

$$T = k\varphi I \qquad (11.2)$$

where φ is the total flux passing through the coil and k is the motor constant that depends on the number of magnetic poles used. $k\varphi$ is constant for a particular motor. The torque is proportional to the armature current.

229

As the armature rotates in the magnetic field a voltage is induced by generator action and a voltage V_b appears across its terminals. V_b is the back electromotive force or back emf given by

$$V_b = 2nBlr\omega = nAB\omega \qquad (11.3)$$

$$V_b = k\varphi\omega \qquad (11.4)$$

Applying the above to the circuit of Fig.11.3 gives

$$I = \frac{V_s - V_b}{R_a} \qquad (11.5)$$

$$I = \frac{V_s}{R_a} - \frac{k\varphi\omega}{R_a} \qquad (11.6)$$

Substituting the above expression for current into Equ. 11.2 gives the torque as

$$T = \frac{k\varphi V_s}{R_a} - \frac{(k\varphi)^2 \omega}{R_a} \qquad (11.7)$$

Equations 11.6 and 11.7 show that as the angular velocity of the motor increases both the current drawn by the motor and the torque developed decrease. Equations 11.7 is a straight line relationship between T and ω. The intercepts give the stall torque (at zero speed) and the free running speed of the motor.

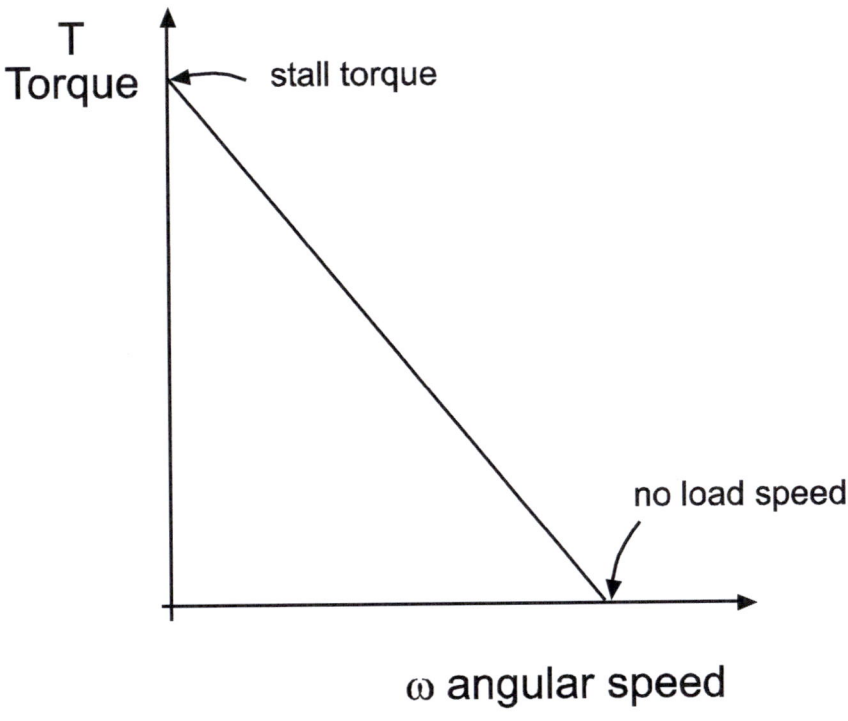

Fig. 11.4 Torque-speed characteristics of pm dc motor

11.4 Synchronous machine

Assume that the permanent magnet (rotor) in Fig. 11.5 rotates at a uniform speed while the stator has three, stationary, coils housed within it so that the plane of each coil is at 120° to the other two. The change in the flux linking each coil induces a sinusoidally varying voltage across the terminals of each coil where the voltages are 120° out of phase with each other.

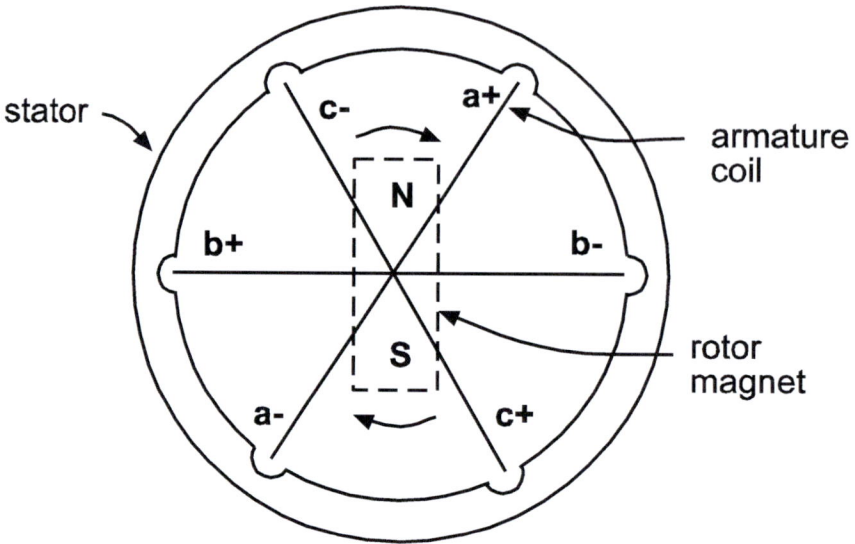

Fig. 11.5 Synchronous machine

For the two pole generator of Fig. 11.5, the the voltage across a coil goes through a complete cycle as the magnet goes through a complete revolution. This synchronization of voltage frequency with speed of rotation characterizes the generator as synchronous. Synchronous generators have a higher power to mass ratio and operate more smoothly than single phase generators.

If each armature coil is now fed an externally generated ac voltage, so that each voltage is 120° out of phase with the other two, then the magnet rotates and the machine acts as a synchronous motor.

11.5 Three-phase circuit

Electrical energy produced by three-phase synchronous generators is transmitted and distributed over three-phase circuits. In addition to the advantages mentioned above, three-phase circuits make more efficient use of wires. Three-phase systems are wye or delta connected. A wye-connected circuit is shown in Fig. 11.6. Here one end of each armature coil is connected to neutral or to ground. The generator voltage that appears between the other terminal (line) of a coil and neutral is the phase voltage. The voltage between lines is a line-to-line, or line, voltage.

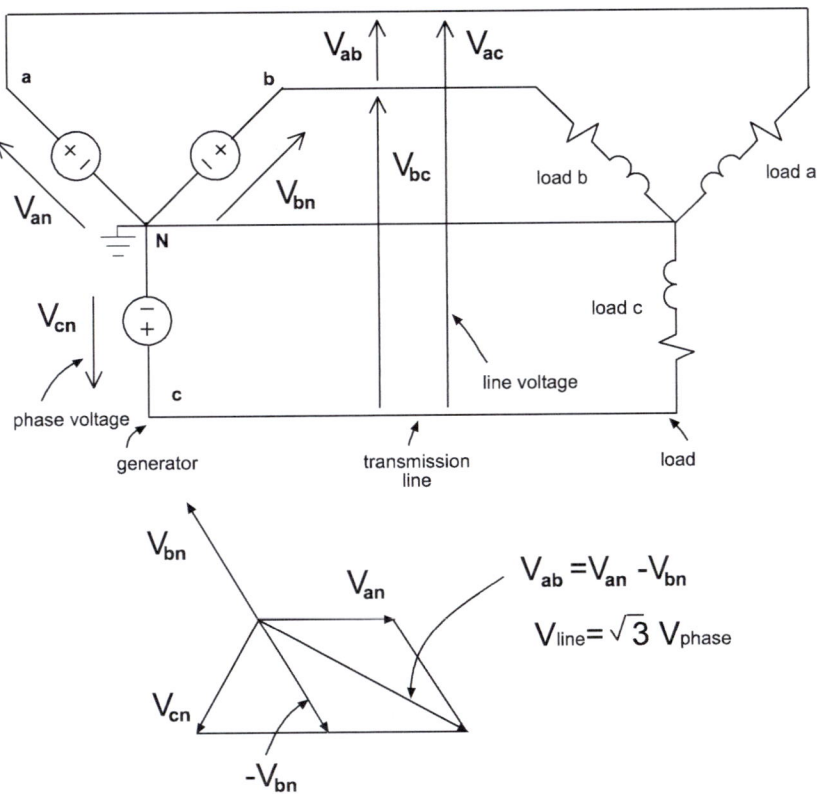

Fig. 11.6 Wye-connected circuit

In a balanced circuit the magnitudes of phase voltages are equal and the phase voltages are given by

$$V_{an} = V_{bn} = V_{cn} = V_a = V_b = V_c = V \tag{11.8}$$

$$\underline{V}_a = V\angle 0°, \quad \underline{V}_b = V\angle 120°, \quad \underline{V}_c = V\angle 240° \tag{11.9}$$

The line voltages are

$$\underline{V}_{ab}, \underline{V}_{bc}, \underline{V}_{ca}$$

From the circuit and phasor diagram of Fig.11.6

$$V_{line} = \sqrt{3} V_{phase} \tag{11.10}$$

The phase current is equal to the line current. If the impedance of each leg of the wye-connected load is the same for each leg, then the load is a balanced load and line currents are all equal in magnitude. One end of each leg of the load is connected to the neutral or ground as shown.

The real power per leg of the load is

$$P_{av} = V_{phase} I_{line} \cos\theta = \frac{V_{line} I_{line} \cos\theta}{\sqrt{3}} \tag{11.11}$$

And for a balanced system the total real power consumed by the load is

$$P_{av} = \sqrt{3} V_{line} I_{line} \cos\theta \ \text{Watts} \tag{11.12}$$

where θ, in the power factor $\cos\theta$, is the phase angle difference between the line current, that passes through each leg of the load, and the phase voltage, since this is the voltage dropped across each leg. Similarly, the total reactive power of the load is

$$Q = \sqrt{3} V_{line} I_{line} \sin\theta \ \text{VAR} \tag{11.13}$$

The apparent power per leg of the load is

$$S = V_{phase} I_{line} = \frac{V_{line} I_{line}}{\sqrt{3}} \ \text{VA} \tag{11.14}$$

Example 11.1 Three-phase power

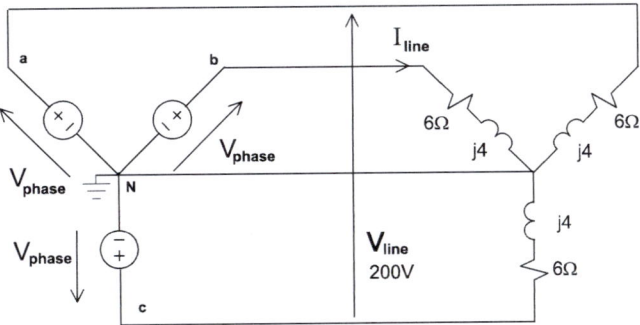

Fig. 11.7 Circuit for Example 11.1

Find the power associated with the load. The line voltage is 200 Vrms.

There are three identical loops and the phase voltage is dropped across a leg of the load in each.

$$I_{line} = \frac{V_{phase}}{6+j4} = \frac{V_{line}}{\sqrt{3}(6+j4)} = \frac{200}{\sqrt{3}(6+j4)} = 16.01\angle -33.69°$$

Phasor power is

$$\underline{P} = 3(\frac{V_{line}I_{line}}{\sqrt{3}}\cos\theta + j\frac{V_{line}I_{line}}{\sqrt{3}}\sin\theta)$$

$$\underline{P} = 3(1848.68\angle 33.69°)$$

$$\underline{P} = 3(1848.68\cos 33.69 + j1848.68\sin 33.69)$$

$$\underline{P} = 3(1538.19 + j1025.46)$$

The total apparent power is 5546.04VA, the total real power is 4614.57W and the total reactive power is 3076.38VAR.

11.6 Photovoltaics

Solar cells convert solar radiation into electrical energy. Before introducing the electrical model of the solar cell some ideas concerning electrical conduction and photoconduction will be discussed.

11.6.1 Electrical conduction

The passage of electric charge through a material is called electrical conduction. Electrical conduction causes an electric current to flow. The current is driven by an electric field and depends on the field, the density of free charge and the ease with which the charge moves in the particular material. The current flowing through a conductor or a resistor, both with some finite resistance to the flow of current, is described by a form of ohm's law as

$$J = \sigma E \tag{11.15}$$

where $J=I/A$ if I is the current flowing at right angles to a cross-sectional area of A m^2 of the conductor, and σ is the electrical conductivity of the material in mho/m (Fig. 11.8). The electric field E (units V/m) is proportional to the voltage difference V applied across the conductor so that if this voltage is dropped across a length d of the conductor then

$$E = \frac{V}{d} \tag{11.16}$$

Flow of electrical current in a metal is due to the directed movement of electrons within the lattice of the metal. In the absence of an electric field electrons in the material move around due to their thermal energy, however, because this movement is random, no current results. The application of the voltage difference between the electrodes on either side of the material, and the subsequent establishment of the electric field within the material, results in a force of magnitude qE on an electron. Electrons in the material now have a net drift velocity and move to the positive electrode, and as long as the negative electrode acts as an ohmic contact to replenish the electrons leaving the material, a current will flow in a closed circuit. The flow of conventional current is in the opposite direction to the flow of electrons (Fig. 11.8).

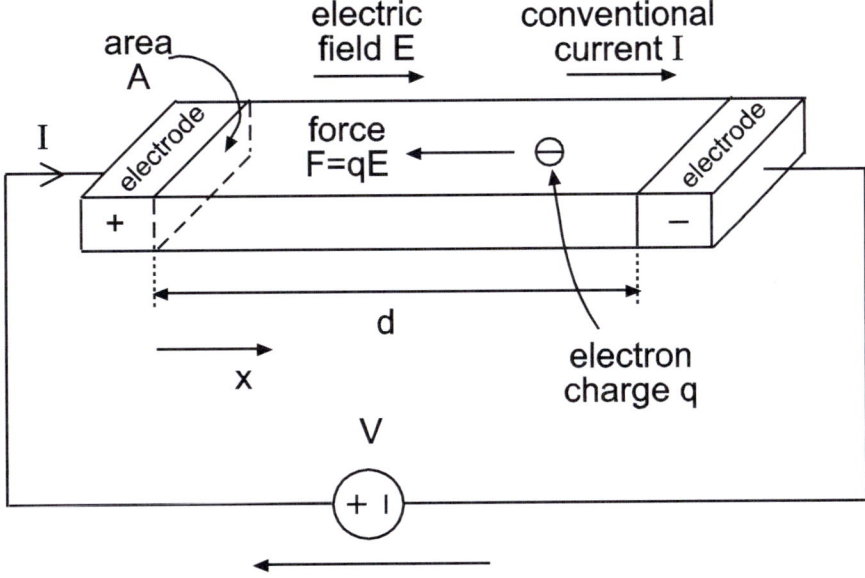

Fig. 11.8 Drift of charge in an electric field

The electric field points in the direction that a positive charge would move if it were free to do so. Also, if x indicates direction

$$E = -\frac{dV}{dx} \qquad (11.17)$$

and the electric field points in the direction of decreasing electrical potential, or voltage, as shown in Fig. 11.8.

237

As the force qE acts on a free electron in the field E, the velocity of the electron increases until it collides with an atom within the crystal lattice of the conductor when it loses the energy it had previously gained. It then accelerates until it is stopped once again and so forth. The end result is that the electron moves through the lattice in the direction opposite to that of the electric field with an average velocity called the drift velocity. This passage of electric charge, or conduction, causes an electric current that is referred to as the drift current. If the average, or drift velocity of the electron is v, the average time between collisions is τ, q is the electronic charge (1.6×10^{-19} C), and m_e is the effective mass of the electron in the material, then from Newton's law

$$qE = m_e \frac{v}{\tau} \qquad (11.18)$$

The ease with which the electron drifts through a particular material is the mobility μ (unit m^2 / V-s) where

$$v = \mu E \qquad (11.19)$$

Eliminating E between equs. 11.18 and 11.19 gives

$$\mu = \frac{q\tau}{m_e} \qquad (11.20)$$

Introducing μ into equ. 11.15 now gives

$$J = nq\mu E \qquad (11.21)$$

where n is the density of conduction electrons (m^{-3}) that are free to move about in the material under the action of the electric field. The conductivity is then

$$\sigma = nq\mu \qquad (11.22)$$

238

The conductivity of a material is the reciprocal of its resistivity ρ

$$\sigma = \frac{1}{\rho} \qquad (11.23)$$

The resistance of the material of length d and cross-sectional area A in Fig. 11.8 is

$$R = \rho \frac{d}{A} \qquad (11.23)$$

From equ. 11.23 the unit of resistivity is the ohm-m. Therefore, conductivity is measured in units of ohm^{-1}m^{-1} or mho/m. The conductivity of conductors such as copper, silver and gold is of the order of 10^7 mho/m while that of good insulators can be up to 10^{-17} mho/m.

11.6.2 Electrical conduction in semiconductors

Conduction in semiconductors is due to the flow of negatively charged electrons and positively charged holes. Figure 11.9 shows the energy band diagram of a semiconductor with allowed energy levels in the conduction and valence bands. The band diagram shows the electron potential energy as a function of distance into the semiconductor bulk. No energy levels exist for electrons to occupy in the forbidden energy band. The forbidden energy band is also called the band gap. At low temperature the valence band is full of electrons and the conduction band is empty of electrons. At room temperature some electrons have enough energy to break away from their parent atoms and be elevated from the valence to the conduction band. The resulting absence of an electron in the valence band constitutes the presence of a positive hole in the valence band. The generation of electron-hole pairs does not result in an ever increasing density of free electrons and holes. This is because   the generation of electron-hole pairs coexists with the

239

recombination of electron-hole pairs so that under thermal equilibrium (ie with the semiconductor in the dark and without any externally applied electric field) a constant density of n electrons and p holes is established in the conduction and valence bands, respectively. Electrons can move in the conduction band while holes can move in the valence band, under the action of an electric field, to give rise to a current. The current is made up of an electron and a hole component. Holes drift in the direction of the electric field while electrons move in the opposite direction. However, because electrons are negatively charged, the electron component of current is in the same direction as the hole component.

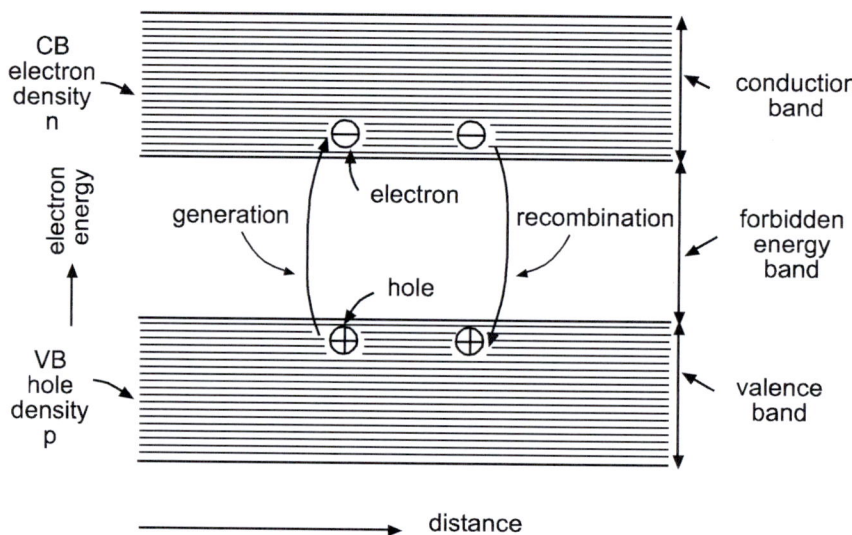

Fig. 11.9 Energy band diagram of semiconductor

If μ_n and μ_p are the electron and hole mobilities, respectively, then the current density through the semiconductor is given by

$$J = q(n\mu_n + p\mu_p)E \tag{11.24}$$

240

and the conductivity of the semiconductor is

$$\sigma = qn\mu_n + qp\mu_p \qquad (11.25)$$

In an intrisic (pure, undoped) semiconductor the density of free electrons equals the density of holes $n = p = n_i$ where n_i is the intrinsic density of free carriers. The width of the forbidden energy band, or the width of the band gap, is 1.12 eV (electron volts) for Si and 0.72 eV for Ge. For silicon at room temperature (300 K), $n_i \approx 1.5 \times 10^{10}$ cm^{-3} and for germanium $n_i \approx 2.4 \times 10^{13}$ cm^{-3}.

Doping intrinsic silicon by introducing a pentavalent element such as phosphorus or arsenic, as a substitional impurity, into the lattice of silicon (tetravalent), results in an n-type semiconductor. Doping silicon with a trivalent element, such as boron or aluminium, results in a p-type semiconductor. Doped semiconductors are extrinsic materials. In extrinsic semiconductors the relationship $np = n_i^2$ holds as long as the semiconductor is in thermal equilibrium, ie it is not subjected to external stimuli.

In n-type silicon a donor energy level exists in the forbidden energy band close to the bottom of the conduction band. At room temperature the electrons in the donor level have enough energy to break away from their parent atoms and to rise to the conduction band. The density of electrons n in the conduction band is now referred to as the majority carrier concentration while the density of holes in the valence band p is the minority carrier concentration since, as $np = n_i^2$, then $n \gg p$.

In p-type silicon an acceptor level exists in the forbidden energy band that can accept electrons from the valence band. This level is located just above the top of the valence band. At room temperature electrons in the valence band have enough energy to move up and occupy this level leaving holes in the valence band. The semiconductor equation $np = n_i^2$ is valid and so $p \gg n$. The holes are now the majority carriers while the electrons in the conduction band are the minority carriers.

241

11.6.3 Photoconduction

The density of free carrirs in a semiconductor can be increased by photoexcitation. When light is shone on the semiconductor, and if the energy of photons impinging on the surface of the semiconductor equals or exceeds the band gap energy, electron-hole pairs are created. An electron-hole pair is created when an electron in the valence band absorbs the energy from a photon and is excited up to the conduction band while leaving a hole in the valence band. A steady state situation is reached when the rate of creation of electron-hole pairs equals the rate of annihilation of electron-hole pairs through recombination. Between creation and annihilation the excess carriers are free for a lifetime of τ seconds (excess carrier lifetime). The process creates an excess density of electrons δn in the conduction band and an excess density of holes δp in the valence band (where $\delta n = \delta p$) which add to the equilibrium concentrations of free electrons n and holes p to increase the conductivity.

If the source of optical excitation is removed from say an n-type semiconductor, then the excess carrier concentration decays exponentially so that if $\delta p(0)$ is the steady-state density of minority carriers just before the light is extinguished at time $t=0$, then for $t > 0$ the excess carrier concentration at time t is

$$\delta p(t) = \delta p(0) e^{-t/\tau_p} \qquad (11.26)$$

where τ_p is the excess carrier hole lifetime. Under steady state conditions,

$$\delta p = G \tau_p \qquad (11.27)$$

where G is the generation rate of free carriers. Assuming G is constant throughout the material then its units will be $m^{-3}\text{-}s^{-1}$.

The excess carriers result in an increase in conductivity, called the photoconductivity $\delta\sigma = q\delta p(\mu_n + \mu_p)$. The photoconductivity when added to the equilibrium conductivity to give the conductivity of the semiconductor under illumination as

$$\sigma = q(n\mu_n + p\mu_p) + q\delta p(\mu_n + \mu_p) \tag{11.28}$$

If the current density passing through the semiconductor in the dark, with an applied electric field E, is J_q, then under illumination the current has the added photocurrent J_p and is given by

$$J = J_q + J_p \tag{11.29}$$

$$J = q(n\mu_n + p\mu_p)E + q\delta p(\mu_n + \mu_p)E \tag{11.30}$$

From Equ 11.27 the total current is then

$$J = q(n\mu_n + p\mu_p)E + qG\tau_p(\mu_n + \mu_p)E \tag{11.31}$$

11.6.4 Solar cell

Photoconductive materials are used in a pn structure to make a solar cell. Fig. 11.10(i) shows the symbol for a pn diode, or a pn junction solar cell, in the dark. Forward biasing the diode, by applying a voltage across the device so that the p-side is positive with respect to the n-side, results in the forward current while in reverse bias the reverse saturation current I_0 flows through the device ($I_0 \approx 10^{-12}$ A/cm^2). In both forward and reverse bias the diode current is given as a function of the applied voltage V by

$$I_d = I_0[\exp(\frac{qV}{kT}) - 1] \tag{11.32}$$

where k is Boltzmann's constant and T the absolute temperature.

At room temperature (300K) and with $k = 1.38 \times 10^{-23}$ J/K
Equ. 11.32 becomes

$$I_d = I_0 [\exp(\frac{V}{0.026}) - 1]$$

(11.33)

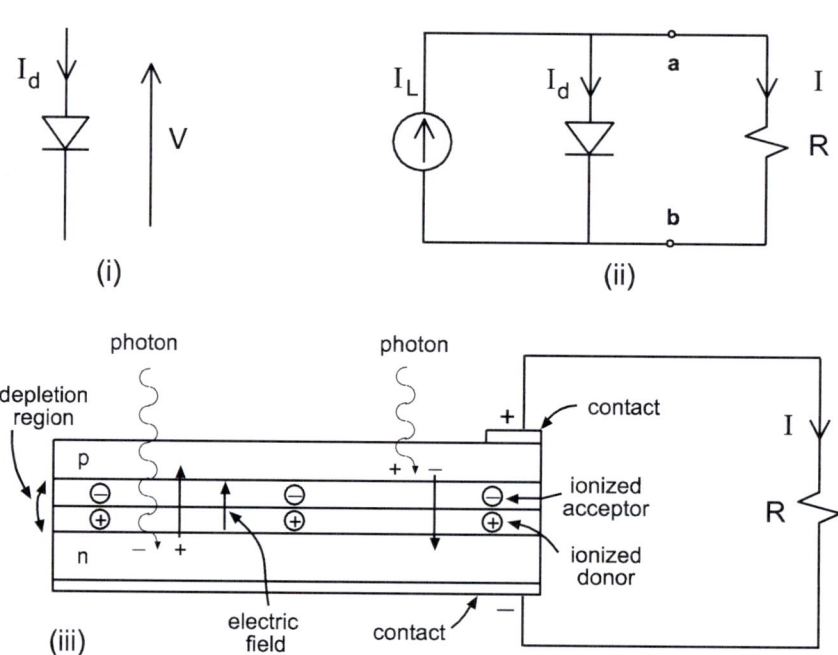

(i)

(ii)

(iii)

Fig. 11.10 (i) Symbol of a pn diode, or of a solar cell, in the dark
(ii) electrical model of a solar cell under illumination, shown to the
left of terminals *a* and *b*, providing power to a load resistor *R* (iii)
the creation of electron-hole pairs by incident photons and the
separation of charge by the electric field at the pn junction of a
solar cell under illumination

The electrical model of the solar cell under illumination is given in Fig. 11.10(ii). The model appears to the left of terminals a and b and consists of an ideal current source of strength I_L in parallel with a diode. The solar cell is shown providing a current to the load resistor R. The net current I flows out of the p-side of the cell and into the load. Under illumination electron-hole pairs are created in the photoconductors on either side of the metallurgical junction of the solar cell (Fig. 11.10(iii)). This creates excess carriers on both sides of the junction and minority carriers on either side of the junction diffuse[+] from regions of high to regions of low density. Minority carriers that reach within a diffusion length of the depletion region[++] at the junction are swept across the junction and accumulate as a net charge on the opposite side of the junction. This results in a net positive charge of holes on the p-side and a net negative charge on the n-side of the pn junction. The device is now a photovoltaic generator of electrical energy or a solar battery.

From Fig. 11.10 (ii) the load current is

$$I = I_L - I_d \tag{11.34}$$

Footnotes:
[+] *The diffusion length L is the average distance that a minority carrier diffuses before it recombines with majority carrier and the average time that it exists as a free carrier is its lifetime* τ, *where* $L = (D\tau)^{1/2}$, $D/\mu = kT/q$ *and D is the diffusion coefficient.*
[++]*A built in electric field exists at a pn junction. This field extends across a region that is depleted of mobile carriers; the depletion region. Immobile, fixed charges exist in the depletion region. Ionized donors on the n-side and ionized acceptors on the p-side provide the positive and negative charges in the depletion region. Since these net charges are separated in space an electric field exists between them. This electric field is responsible for the separation of the photogenerated minority carriers and the subsequent appearance of the voltage across the solar battery.*

and substituting for the diode current from Equ. (11.32)

$$I = I_L - I_0[\exp(\frac{qV}{kT}) - 1]$$
(11.35)

With the terminals of the solar cell short circuited (Fig. 11.11 (i)) the voltage across the diode is zero and Equs. 11.34 and 11.35 give

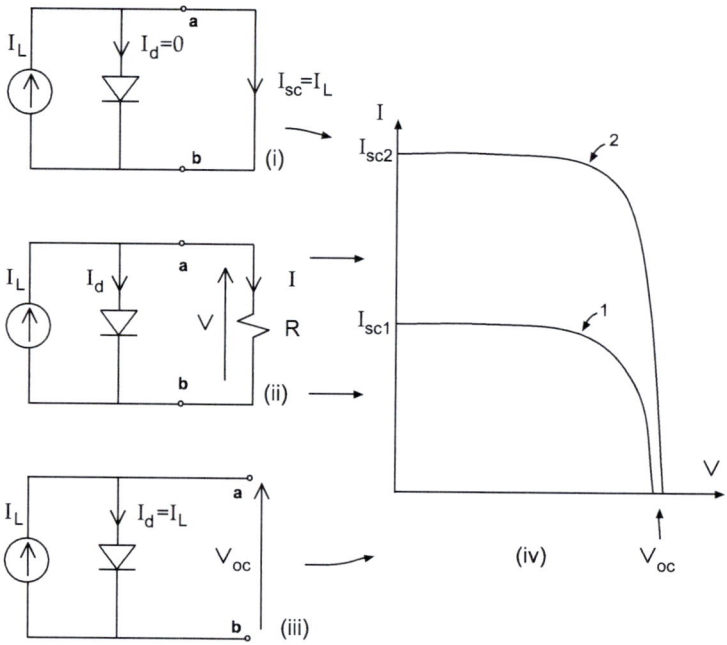

Fig. 11.11 solar cell current-voltage characteristics in (iv). I_{sc}=short circuit current, V_{oc}=open circuit voltage. $I_{sc2} = 2I_{sc1}$ since the insolation for curve 2 is assumed to be twice that for curve 1.

$$I_L = I_{sc}$$
(11.36)

$$I = I_{sc} - I_0[\exp(\frac{qV}{kT}) - 1]$$
(11.37)

where I_{sc} is the short circuit current.

With the solar cell terminals on open circuit (Fig. 11.11 (iii)) $V=V_{oc}$ and $I_{sc}=I_d$ and

$$I_{sc} = I_0[\exp(\frac{qV}{kT}) - 1]$$
(11.38)

$$V_{oc} = \frac{kT}{q}\ln(\frac{I_{sc}}{I_0} + 1)$$
(11.39)

The short circuit current is also given by

$$I_{sc} = I_L = qG(L_n + L_p)A$$
(11.40)

where L_n and L_p are the diffusion lengths of electrons and holes, respectively, and A is the area of the solar cell that is under illumination. Equation 11.40 shows that I_{sc} is directly proportional to the generation rate[+] G of electron-hole pairs which in turn is proportional to the incident solar radiation (insolation). Equation 11.37 is the current-voltage relationship for a photovoltaic cell. The plot of the *I-V* characteristics of a solar cell is shown in Fig. 11.11(iv) where $I_{sc2} = 2I_{sc1}$ since the insolation for curve 2 is assumed to be twice that for curve 1.

AM0 (air mass ratio zero) illumination refers to extraterrestrial insolation where there is no atmosphere to absorb the solar energy. AM1 radiant energy is the sun's energy at sea level under a clear sky with the sun directly overhead. AM1.5 insolation is used as the average solar energy at the earth's surface.

Footnote:

[+] *A photoconductor of bandgap $E_g=1.58$ electron volts (eV), when illuminated by photons of wavelength $\lambda = 1.24/E_g = 1.24/1.58 = 0.78\mu m$, at a rate of 500W/m² or 500J/m²-s will have $G=500/(1.58 \times 1.6 \times 10^{-19}) = 1.98 \times 10^{21}$ electron-hole pairs created within its volume if each incident photon creates an electron-hole pair and if the absorption is uniform throughout the material. In practice the illumination rate should be multiplied by the absorption coefficient α (cm⁻¹). An electron volt is the energy acquired by an electron that moves through a potential drop of 1 volt.*

A more precise electrical model of the solar cell appears in Fig. 11.12. This includes a series resistance R_s and a shunt resistance R_{sh}. The series resiastance is due to an unusually high semiconductor resistance or due to bad contacts, while the shunt resistor is due to weak regions. The voltage V across the load resistor R_L and the load current I are related through the equation

$$I = I_L - I_0 \{\exp[\frac{q(V + IR_s)}{kT}] - 1\} - \frac{(V + IR_s)}{R_{sh}} \qquad (11.41)$$

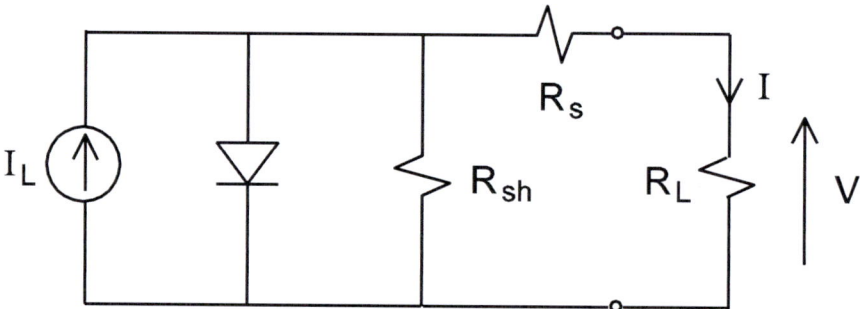

Fig. 11.12 Electrical model of a solar cell that includes the series resistance R_s and the shunt resistance R_{sh} of the cell. The I-V curve refers to the voltage V appearing across the external load resistor R_L and the current I flowing though the load resistor

The shunt resistance determines the slope of the I-V curve of the cell near the short circuit current and the series resistance determines the slope of the I-V curve near the open circuit voltage (Fig. 11.13). This follows from the equivalent resistance of the cell as seen by the load (Fig. 11.14). For small values of the output voltage V the diode resistance R_d is very large and equivalent resistance is $R_{sh} + R_s$, and since $R_{sh} \gg R_s$ the variation of V with I equals R_{sh}. For larger values of V the diode resistance is very small and so the terminals of R_{sh} are shorted and the equivalent resistance is R_s.

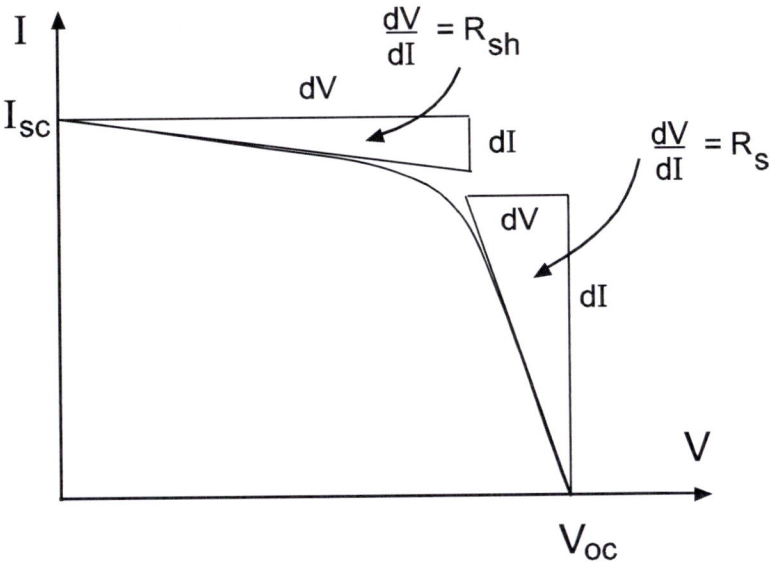

Fig. 11.13 Determining shunt and series resistances from I-V curve
of solar cell

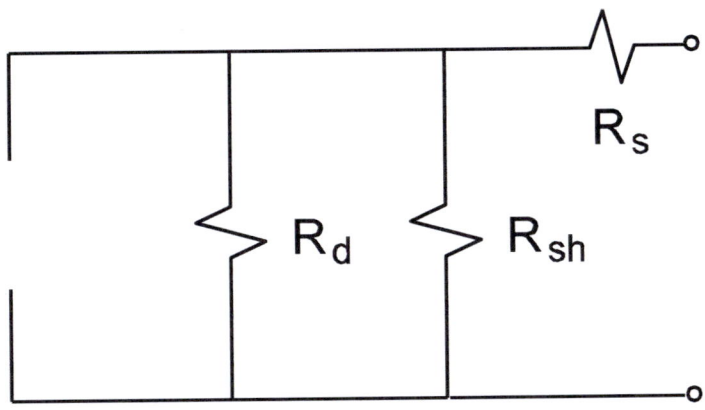

Fig. 11.14 Equivalent resistance of solar cell

249

The energy conversion efficiency of the solar cell is given by $\eta = P_{out}/P_{in} = I_{mp}V_{mp}/P_{in} = V_{oc}I_{sc}FF/P_{in}$ where P_{in} is the optical power input and P_{out} the electrical power output. I_{mp} and V_{mp} are the current and voltage at the maximum power point which occurs at the knee of the I-V relation (Fig. 11.15). FF is the fill factor.

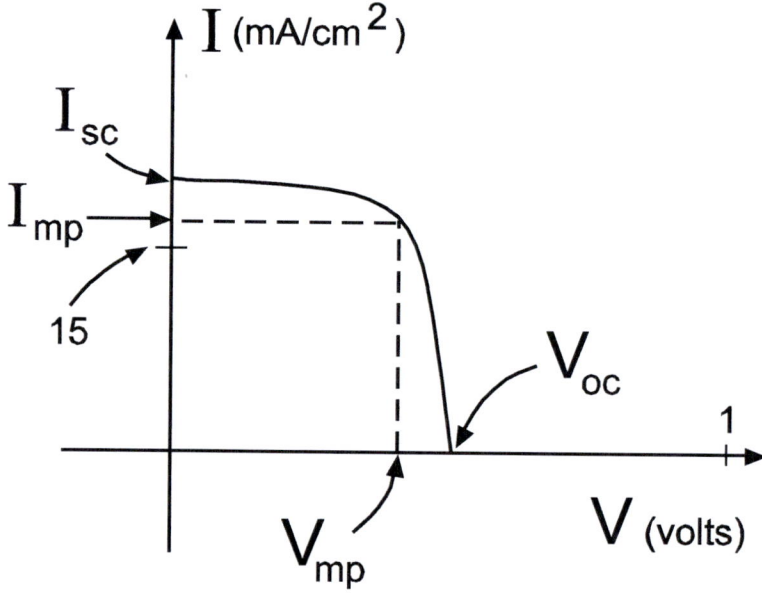

Fig. 11.15 Solar cell current-voltage curve showing the current I_{mp} and voltage V_{mp} at the knee of the curve where maximum power is delivered to the load

11.7 Battery
11.7.1 Primary and secondary batteries

An electric cell produces an electromotive force, or an electric potential difference, that can do work. A cell is a single element of a battery and generates a dc voltage by converting chemical energy into electrical energy. A battery is formed when several cells are connected together. The voltage that appears across the terminals of a battery can drive a discharge current through a resistor, connected across the battery, thereby converting electrical energy into heat. A primary battery is non-rechargeable. A secondary battery must be properly charged before use and it is subsequently rechargeable. A secondary cell converts chemical energy into electrical energy, through a reversible chemical reactions, and is also called a storage cell. Secondary batteries are used to store electrical energy generated by photovoltaic panels. The stored energy is then available during the night or during cloudy days.

11.7.2 Battery as generator or absorber

If a battery of voltage V_b is connected across a resistor in series with a voltage source of voltage V_a so that $V_b > V_a$, as shown in Fig. 11.16(i), or when the battery is connected across the load resistor in Fig. 11.16(ii), the battery is the generator. The current that flows out of the positive terminal of the battery is the discharge current because it starts to deplete the energy stored in the battery.

If an external voltage source of voltage V_g is connected across a secondary battery of terminal voltage V_b so that $V_g > V_b$, then the voltage source acts as a generator and the battery as an absorber with current flowing into the positive terminal of the battery and out of the negative terminal (Fig.11.16(iii)). The current is now the charging current as it induces electrochemical changes within the battery which result in the storage of energy. This results in the recharging of the battery. Lead-acid and nickel-cadmium batteries are secondary batteries.

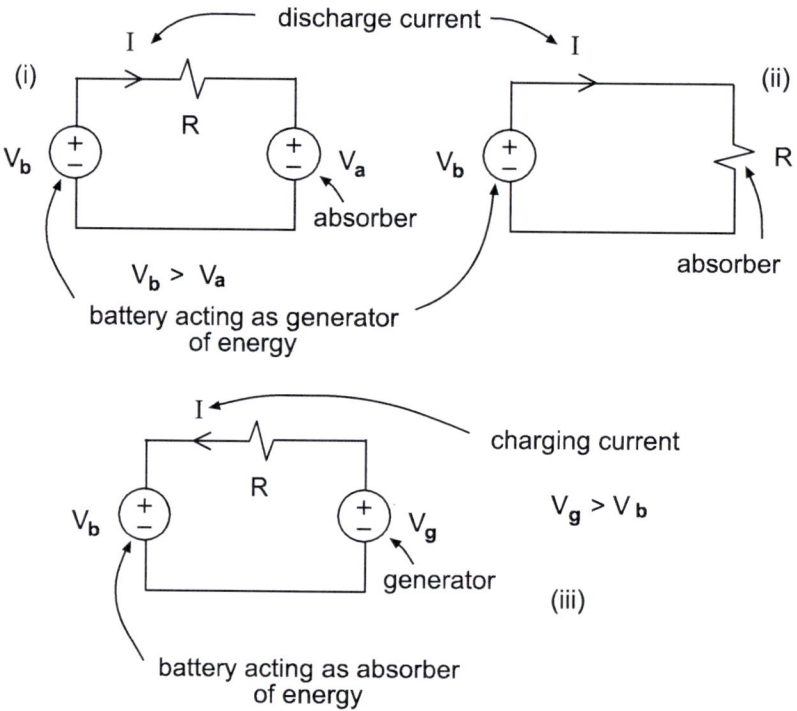

Fig. 11.16 Battery acting as an energy generator in (i) and (ii) and as an absorber in (iii)

11.7.3 Battery capacity, charge and discharge rate

A battery is characterized by its capacity C. The capacity is the the electrical charge that the battery can deliver. Capacity is measured in Ah (ampere-hours) or in mAh (milliampere-hours). The capacity of a battery is determined by discharging a constant current at room temperature ($25°C$) and measuring the time it takes for the voltage to reach a certain low value. The product of the current and the time is the capacity.

Battery specifications give the rated capacity at a specific discharge rate (discharge current). A battery with a rated capacity of $C=2.5Ah$ for a 10 hour rate (C/10 rate), will deliver 0.25A for

10 hours at room temperature. The capacity depends on the discharge rate. A discharge rate higher than the rated one lowers the capacity. A rate lower than the rated one results in a higher capacity. If the discharge rate of a battery is $C/10$ then for two fully

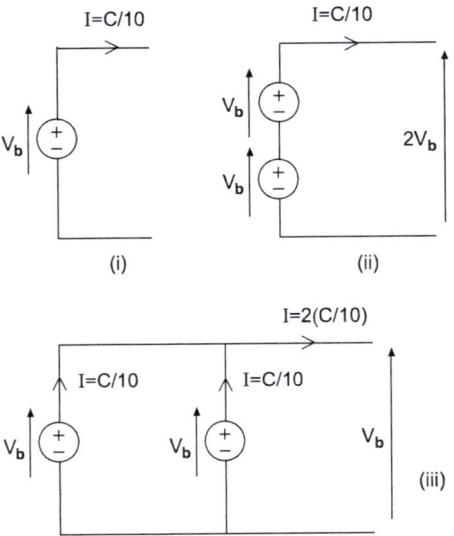

Fig. 11.17 Discharge rate for (i) single battery,
and for batteries connected in (ii) series (iii) parallel

charged identical batteries connected in series, the system has a rating of C Ah, same as for an individual battery, and is capable of a 10 hour discharge rate of $C/10$. The terminal voltage is twice the voltage of an individual battery (Fig. 11.17(ii)). Two identical batteries, each with a nominal terminal voltage of V_b, connected in parallel have a system rating of $2C$ and can provide a 10 hour discharge rate of $2(C/10)$ at a nominal terminal voltage of V_b (Fig. 11.17(iii)).

11.7.4 Basic battery model

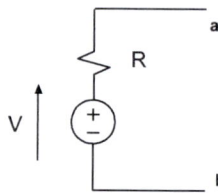

Fig. 11.18 Basic model of battery

A basic model of a battery is shown in Fig. 11.18. It consists of an ideal voltage source of voltage V in series with a resistor R, the internal resistance of the battery. V is the voltage that appears across the battery terminals a and b when these are on open circuit and is therefore often represented by the symbol of the open circuit voltage V_{oc}.

11.7.5 Electrical equivalents of battery combinations

Fig. 11.19(i) shows n batteries of different open circuit voltages and internal resistances connected in series. The system can be represented by the Thévenin equivalent circuit of Fig. 11.19(ii). V_{Th} is the voltage across a and b, in the circuit of Fig.11.19(i), with these terminals on open circuit where

$$V_{Th} = V_1 + V_2 + ... + V_n \qquad (11.42)$$

Then looking into a and b with the voltage sources deactivated gives the Thévenin resistance as

$$R_{Th} = R_1 + R_2 + ... + R_n \qquad (11.43)$$

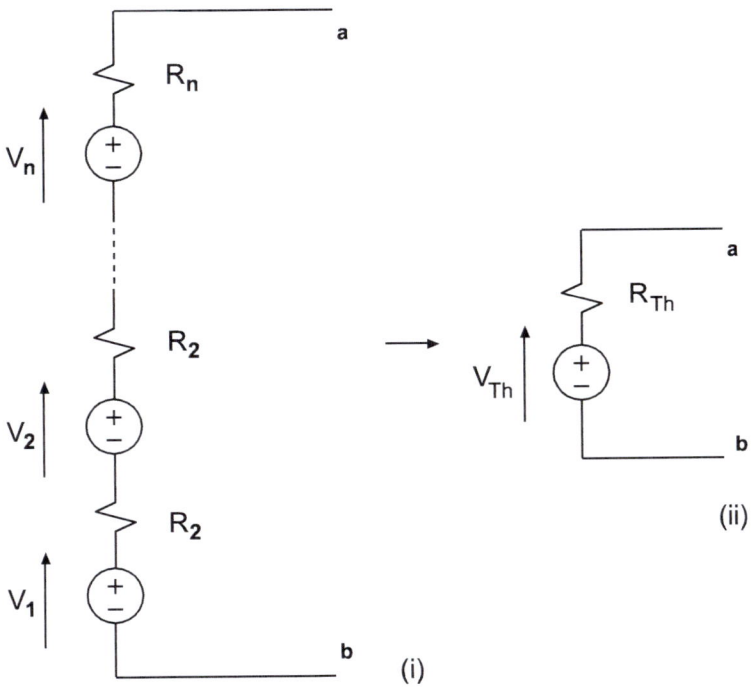

Fig. 11.19(i) Batteries connected in series (ii) the Thévenin
equivalent circuit

In Fig.11.20(i) n batteries of unequal open circuit voltages are
connected together in parallel. The aim is to find the Thévenin
equivalent circuit between terminals a and b (Fig.11.20(i)).
Deactivating the voltage sources in Fig. 11.20(i) leaves the internal
resistors in parallel and so the Thévenin resistance is

$$R_{Th} = \frac{1}{1/R_1 + 1/R_2 + ... + 1/R_n} \qquad (11.44)$$

255

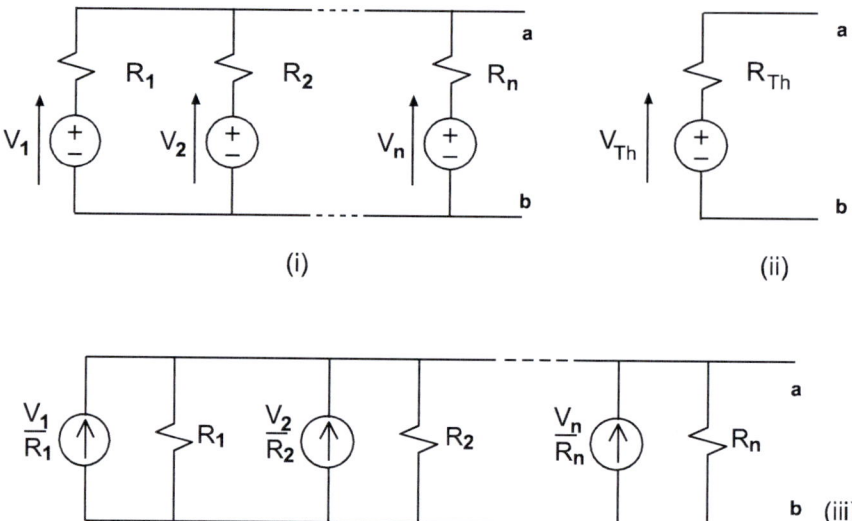

Fig. 11.20(i) Batteries connected in parallel (ii) the Thévenin equivalent circuit (iii) batteries replaced by Norton equivalents

Each battery equivalent of Fig. 11.20(i) is then replaced with its Norton equivalent to get the circuit of Fig. 11.20(iii). The Norton equivalent current of the circuit to the left of *a* and *b*, in Fig. 11.20(iii), is the short circuit current between *a* and *b*. This is given by

$$I_N = \frac{V_1}{R_1} + \frac{V_2}{R_2} + ... + \frac{V_n}{R_n} \qquad (11.45)$$

The Thévenin voltage is then given by

$$V_{Th} = I_N R_{Th} \qquad (11.46)$$

$$V_{Th} = \left(\frac{V_1}{R_1} + \frac{V_2}{R_2} + ... + \frac{V_n}{R_n} \right)\left(\frac{1}{1/R_1 + 1/R_2 + ... + 1/R_n} \right) \qquad (11.47)$$

256

11.7.6 Battery efficiency

The efficiency of a secondary battery is the ratio of the generated energy to the absorbed energy. The discharge and charging currents flow during the generator and absorber phases, respectively. Both charging and discharging currents are expressed in terms of C/t rates.

The total energy absorbed during charging U_a and that generated during discharge U_g are

$$U_a = \int_0^{t_c} v_c(t) i_c(t) dt \qquad (11.48)$$

$$U_g = \int_0^{t_d} v_d(t) i_d(t) dt \qquad (11.49)$$

where i_c and i_d are the charging and discharging currents, v_c is the externally applied voltage during charging, v_d the battery voltage during discharge, and t_c and t_d are the charging and discharging times, respectively. If the charging and discharging currents are approximately constant at I_c and I_d, and the external charging voltage and the battery voltage during discharge, V_c and V_d, are also constant, then

$$U_a = V_c I_c t_c \qquad (11.50)$$

$$U_g = V_d I_d t_d \qquad (11.51)$$

The energy efficiency of the battery is then

$$\frac{U_g}{U_a} = \frac{V_d I_d t_d}{V_c I_c t_c} = \left(\frac{V_d}{V_c}\right)\left(\frac{I_d t_d}{I_c t_c}\right) \qquad (11.52)$$

where (V_d/V_c) is the voltage efficiency and the ratio of electric charges ($I_d t_d / I_c t_c$) is the coulomb efficiency of the battery. The generated energy term $V_d I_d t_d$ in Equ. 11.51 is in units of watts-hours or ampere-hour capacity times the average discharge voltage. Dividing the generated energy by the battery weight or volume gives the energy density that the battery is capable of generating. These densities will be in units of watts-hours/gram or watts-hours/cm^3. The charging and discharging times are decided by the choice of the end-of-charge and the end-of-discharge voltages. These are the respective battery voltages at the end of the charge and the end of the discharge periods.

Chapter 12
Feedback

12.1 Introduction

An amplifier has negative feedback when a portion of the output signal is fed back to oppose the input signal. The gain of the amplifier can be made stable by using negative feedback. In the case of opamps, for instance, the gain of the amplifiers are set by component values external to the amplifier so that the gain is minimally affected by variations in properties of the individual opamp ICs. Negative feedback keeps the amplifier characteristics linear. Increase in the input signal without negative feedback can result in nonlinear amplifiers with resulting distortions in the output signal. Negative feedback in the inverting and non-inverting opamp cases will be considered in this chapter.

12.2 Feedback amplifier equation

The negative feedback system in its general form is given in Fig. 12.1. The gain of the network without feedback is A_o, so that if V_{in} and V_o are the input and output voltages, then

$$V_o = A_o V_{in} \tag{12.1}$$

A_o is called the open loop gain.

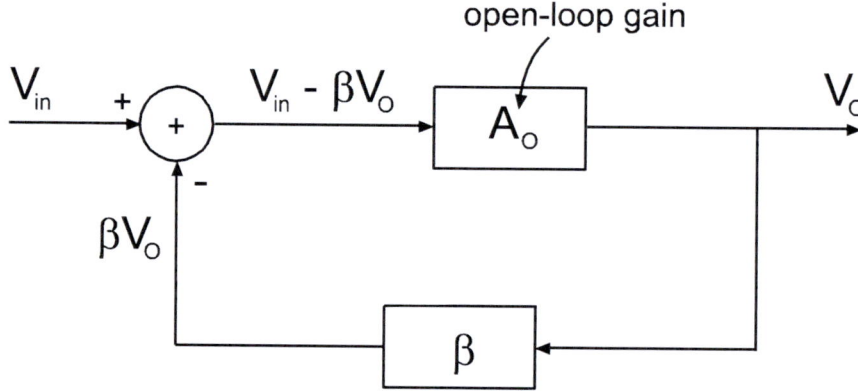

Fig. 12.1 Negative feedback network

If β is the fraction of the output signal that is fed back to the summer to be subtracted from the input signal then the output becomes

$$V_o = A_o (V_{in} - \beta V_o) \tag{12.2}$$

and therefore the closed loop gain is

$$A_c = \frac{V_o}{V_{in}} = \frac{A_o}{1 + \beta A_o} \tag{12.3}$$

where βA_o is called the loop gain. When the loop gain is much larger than unity the closed loop gain is given by

$$A_c = \frac{V_o}{V_{in}} = \frac{1}{\beta} \qquad (12.4)$$

This result means that with negative feedback applied in the network the gain depends on components in the feedback loop. This is used to stabilize the gain of the amplifier. There are limits to the amount of loop gain that can be used. This is because of phase shifts within the loop that for certain frequencies result in positive feedback that can lead to oscillations and instability.

Example 12.1 Amplifier gain stabilization

Two amplifiers have open loop gains of 1000 and 5000. What is the closed loop gain when negative feedback is used with $\beta=0.1$?

For $A_o=1000$ equ. 12.3 gives
$A_c=1000/(1+1000\times0.1)$
$A_c=9.90$

For $A_o=5000$
$A_c=5000/(1+5000\times0.1)$
$A_c=9.98$

Therefore a 400% increase in the amplifier gain without feedback results in a mere 0.8% change in the closed loop gain.

12.3 Inverting opamp

Figure 12.2 shows the inverting opamp circuit with its input resistance R_i carrying a small input current I_{in}. V_s is the source voltage, R_s the source resistance and R_f the resistance in the feedback loop. A_o is the open loop gain of the opamp. The output resistance of the opamp is assumed to be small and has been ignored here.

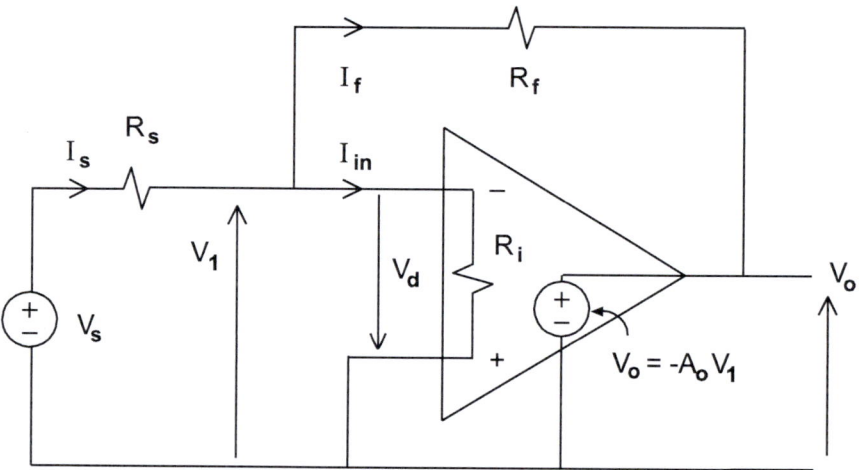

Fig. 12.2 Inverting opamp with negative current feedback

Since the inverting input is at a higher potential than the noninverting one, the output voltage is negative. The output is connected to the inverting input through R_f and the current I_f provides the negative feedback. I_f is subtracted from the source current I_s since from KCL

$$I_{in} = I_s - I_f \tag{12.5}$$

$$\frac{V_1}{R_i} = \frac{V_s - V_1}{R_s} - \frac{V_1 - V_o}{R_f} \tag{12.6}$$

262

Using $V_o = -A_oV_1$, $R_i >> R_s$ and $R_i >> R_f$, equ. 12.6 gives

$$\frac{V_o}{V_s} = -\frac{\left\{A_o\left(\dfrac{R_f}{R_f + R_s}\right)\right\}}{1 + \left(\dfrac{R_s}{R_f}\right)\left\{A_o\left(\dfrac{R_f}{R_f + R_s}\right)\right\}} \tag{12.7}$$

which is of the same form as the general feedback equ. 12.3.

12.4 Non-inverting opamp

The non-inverting opamp uses negative voltage feedback (Fig. 12.3). KCL gives

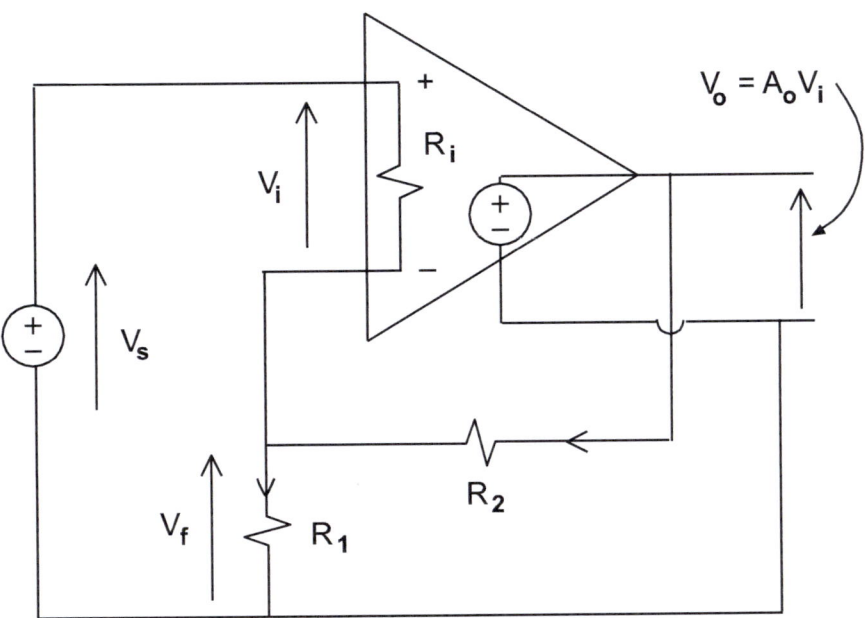

Fig. 12.3 Non-inverting opamp with negative voltage feedback

$$V_i = V_s - V_f \tag{12.8}$$

and since

$$V_o = A_o V_i \tag{12.9}$$

and

$$V_f = \frac{R_1}{R_1 + R_2} V_o \tag{12.10}$$

with

$$\beta = \frac{R_1}{R_1 + R_2} \tag{12.11}$$

$$V_f = \beta V_o \tag{12.12}$$

The relationship between the output voltage and the externally applied source voltage, with the latter acting as the the input, is then

$$\frac{V_o}{V_s} = \frac{A_o}{1 + \beta A_o} \tag{12.13}$$

Chapter 13
Bioamplifier

13.1 Introduction

This chapter describes the electric circuit that can measure the electrical signal of the heart.

The heart is an electrically activated pump. During every cycle of the normal heart, electrical excitation (depolarization) of the atria initiates contraction of the atria which then push blood into the ventricles. This is followed by the ventricles contracting (depolarizing) and pushing blood into the arteries, while the atria relax (electrically repolarize). Electrical activity flows through the nerves that constitute the conducting tract of the heart which then excite the muscle fibres of the heart. Excitation of muscle brings about contraction. For every cycle of the heart electrical activity starts at the sinoatrial node and then spreads to the atrial muscle to bring about atrial contraction. The excitation then travels along the conducting tract to the atrioventricular node, the common bundle, the bundle branches and then to the Purkinje fibres. The Purkinje fibers connect with the muscle fibres of the ventricles which then contract.

Electrical depolarization and repolarization of atrial and ventricular muscle is initiated by a time-dependent current that flows through the membrane that envelops the conducting tract of the heart. This flow of energy is unlike the flow of charge in a closed loop encountered in electrical engineering circuits. In the human heart the membrane that envelops the conducting tract of the heart separates the intracellular fluid from the extracellular (outside) one. In the relaxed (resting) state the intracellular fluid has a net negative charge with respect to the extracellular one. Therefore the voltage inside of the nerve is negative with respect to the outside. The nerve is said to be polarized and the voltage difference is referred to as the membrane potential. Electrical excitation brings about depolarization where the inside starts to become less negative and then positive with respect to the outside.

265

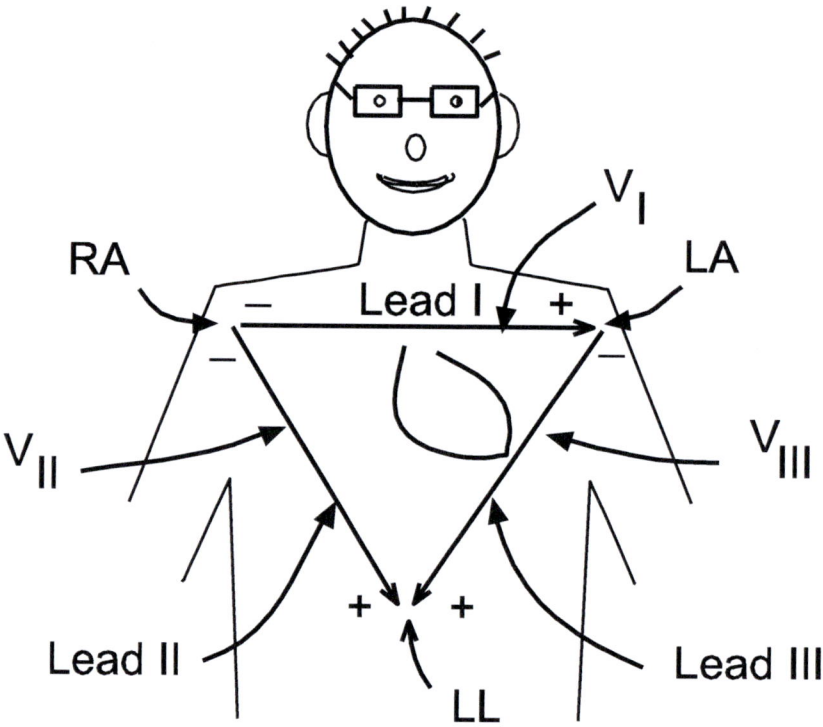

Figure 13.1 The Einthoven triangle representing the electrical activity of the heart.

The membrane then relaxes back to the initial state (repolarizes) with the intracellular fluid once again more negative than the outside. The entire voltage swing in the voltage across the nerve membrane, that accompanies depolarization and then repolarization, is the action potential which starts at a point and then moves along the nerve. In electrically excitable tissue, such as nerve and muscle, action potentials start at one point and travel to another point where the two points are not connected in space as they would be in an electrical engineering circuit.

The electrical activity associated with the contraction and relaxation of cardiac muscle gives a time-dependent distribution of voltage on the human body. This electrical activity is measured in

266

electrocardiography by using body-surface electrodes in a lead system which are then input to a bioamplifier.

A lead, in electrocardiography, is a pair of silver-silver chloride electrodes and the conducting wires connecting these to a bioamplifier. The resulting plot of voltage versus time is the electrocardiogram corresponding to that lead. In the Einthoven lead system, the electrocardiogram (ECG) for lead I gives the voltage of the left arm (LA) with respect to the right arm (RA) and the lead II ECG records the voltage of the left leg (LL) with respect to the RA. The lead III recording is the voltage of the LL with respect to the LA. The electrodes can be attached to the limbs or the chest. The Einthoven representation of the electrical activity of the heart is shown in Fig. 13.1. The heart is within an equilateral triangle with apices at the right and left shoulders and the pubis, corresponding electrically to electrodes at the right arm (RA), the left arm (LA) and the left leg (LL), respectively. The arrows for the three potential differences of Fig. 13.1 form the Einthoven triangle. Voltage differences are measured in three directions; V_I along lead I (RA-LA), V_{II} along lead II (RA-LL), and V_{III} along lead III (LA-LL). The relationship between the three lead voltages V_I, V_{II} and V_{III}, is $V_I = V_{II} - V_{III}$.

A lead I ECG is given in Fig. 13.2. The P wave is due to the atria contracting. The QRS wave corresponds to the ventricles contracting. The QRS complex overshadows the smaller wave due to the atria relaxing since this occurs at the same time as ventricular contraction. The T wave corresponds to the ventricles relaxing.

The voltage difference that develop across the body as a result of cardiac activity is of the order of millivolts. The task of the bioamplifier used to measure this voltage is to amplify this voltage difference while reducing the common mode voltage that is also present at the input to the amplifier.

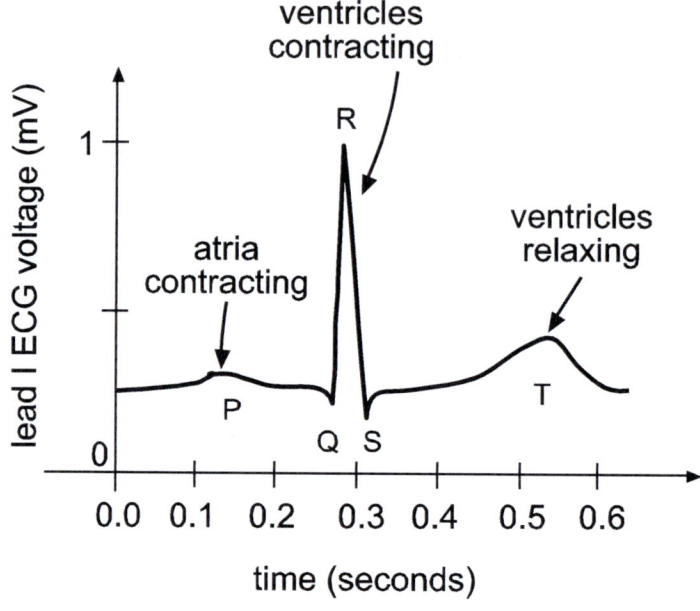

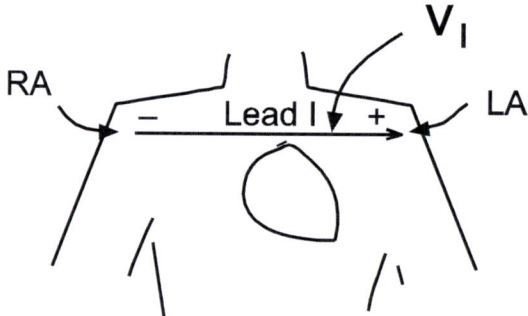

Figure 13.2 A sketch of a lead I electrocardiogram showing the
lead I voltage V_I as a function of time.

13.2 Equation of instrumentation amplifier differential gain

The bioamplifier includes the instrumentation amplifier circuit of Fig. 13.3. Op amps 1 and 2 provide the differential gain that is controlled by the resistor R_g. The instrumentation amplifier comes

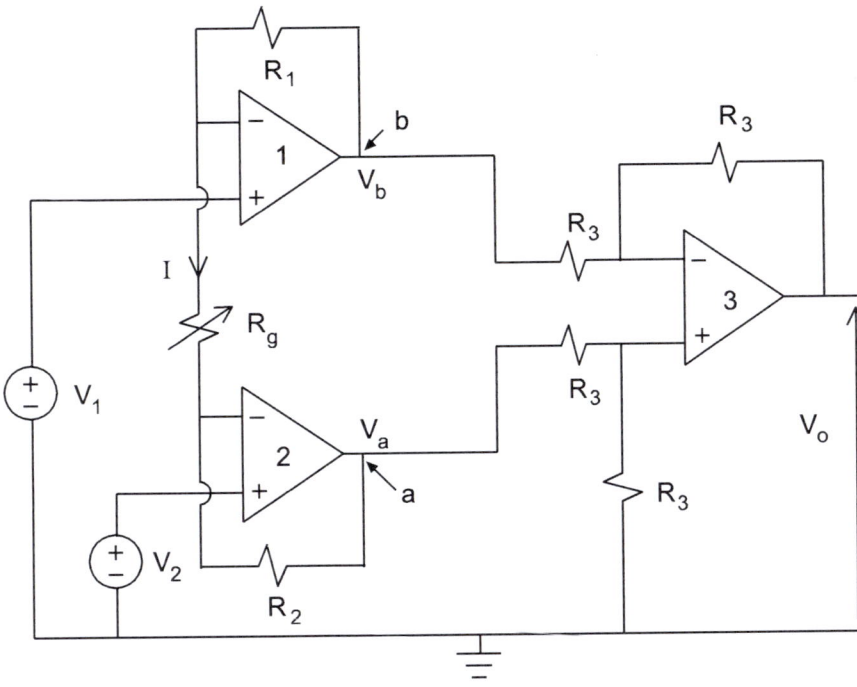

Fig. 13.3 The instrumentation amplifier within the bioamplifier circuit

packaged in a DIP that incloses the three op amps of Fig. 13.3 and all the resistors except R_g. This is very convenient as only one component external to the instrumentation amplifier needs to be chosen. Op amp 3 is part of the unity gain subtractor that is designed to remove the common mode signal. Op amp 3 also converts the floating signal V_{ba} input to it to an output voltage V_o referred to ground.

Let V_l be the left arm voltage and V_2 the right arm voltage where $V_l > V_2$ and $V_l = (V_l - V_2)$ is the lead I voltage. then the differential gain of the amplifier is G where

$$G = \frac{V_b - V_a}{V_1 - V_2} \qquad 13.1$$

Since the inverting inputs to op amps 1 and 2 are V_l and V_2, respectively and assuming that no current enters either input to these op amps, then the current through R_g as well as through R_l and R_2 is

$$I = \frac{V_1 - V_2}{R_g} \qquad 13.2$$

and the voltage difference between b and a is

$$V_b - V_a = V_1 + \frac{R_1 (V_1 - V_2)}{R_g} - \left[V_2 - \frac{R_2 (V_1 - V_2)}{R_g} \right] \qquad 13.3$$

$$V_b - V_a = (V_1 - V_2) \left[1 + \frac{(R_1 + R_2)}{R_g} \right] \qquad 13.4$$

The equ. 13.4 gives the differential gain from inputs to outputs of op amps 1 and 2 as

$$G = \frac{(V_b - V_a)}{(V_1 - V_2)} = 1 + \frac{(R_1 + R_2)}{R_g} \qquad 13.5$$

In the AD620 instrumentation amplifier $R_l = R_2 = 24.7$ kΩ and for this IC

$$G = \frac{49.4 \times 10^3}{R_g} + 1 \qquad\qquad 13.6$$

and G can be set by appropriately choosing R_g in equ. 13.6.

13.3 The power line induced body voltage

The amplifier of Fig.13.3 is a difference amplifier; amplifying the difference between its two inputs while rejecting any signal common to both inputs. An example of the rejected signal would be a noise signal picked up by both input terminals The noise signal would be the common mode voltage.

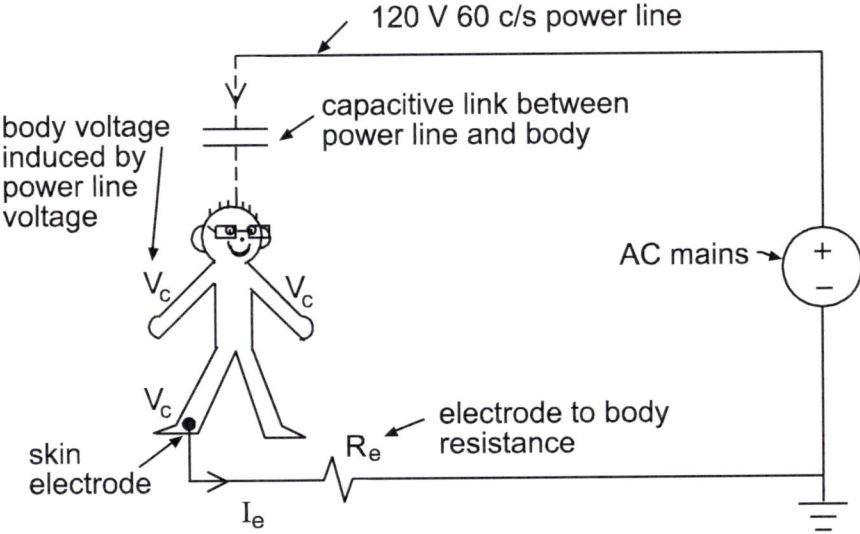

Fig. 13.4 A voltage is induced on the body due to AC power line.

There is a common mode voltage induced on the body because of the proximity of AC power lines in the office environment that electrocardiograms are recorded. Although the induced voltage is the same throughout the skin, the voltages at the two inputs to the amplifier are not identical when two points on the body are attached to the amplifier through skin electrodes and lead wires.

This is because the electrodes that make electrical contact between the lead wires and the skin do not have identical impedances. By the time the common mode signal on the body passes through the electrodes and reaches the amplifier terminals it becomes a differential voltage input to the amplifier of the order of 10 μV. This noise is then amplified by the amplifier and interferes with the cardiac signal.

Practical values of the induced body voltage and capacitive link can be calculated using basic electrical principles. Fig. 13.4 shows how an AC current flows from the AC power line through the capacitive link to the body. In this case air is the dielectric between electrodes and the AC power line and the body are the electrodes. This AC current if of the order of 100 nA (1 nanoampere=10^{-9} amp). Assuming a current of 0.1 μA flowing through the right leg and through the contact resistance of 50 kΩ to the leg and then to ground, means the induced common mode body voltage V_c is 5 mV. Since the same current flows from the 120 V, 60c/s AC mains line to the body, then the capacitace linking the power line to the body is

$$C = \frac{0.1 \times 10^{-6}}{2\pi 60 \times 120} = 2.2 \times 10^{-12} \text{ Farads=2.2 pF (picofarads)} \qquad 13.7$$

The effect of the AC induced body voltage at the output of the amplifier of Fig. 13.3 can swamp the cardiac signal. The way to reduce this undesireable effect is to reduce the magnitude of the induced body voltage. This is done by using the instrumentation amplifier with an additional op amp in a feedback loop.

13.4 Circuit to reduce mains induced body voltage

This circuit is shown in Fig. 13.5. Two resistors, each of value R_4 are connected in series across the gain control resistor R_g. The center of the two R_4 resistors is connected to the feedback op amp 4. Since the non inverting input to op amp 4 is grounded the center node d between resistors R_4 is at virtual ground. Since the

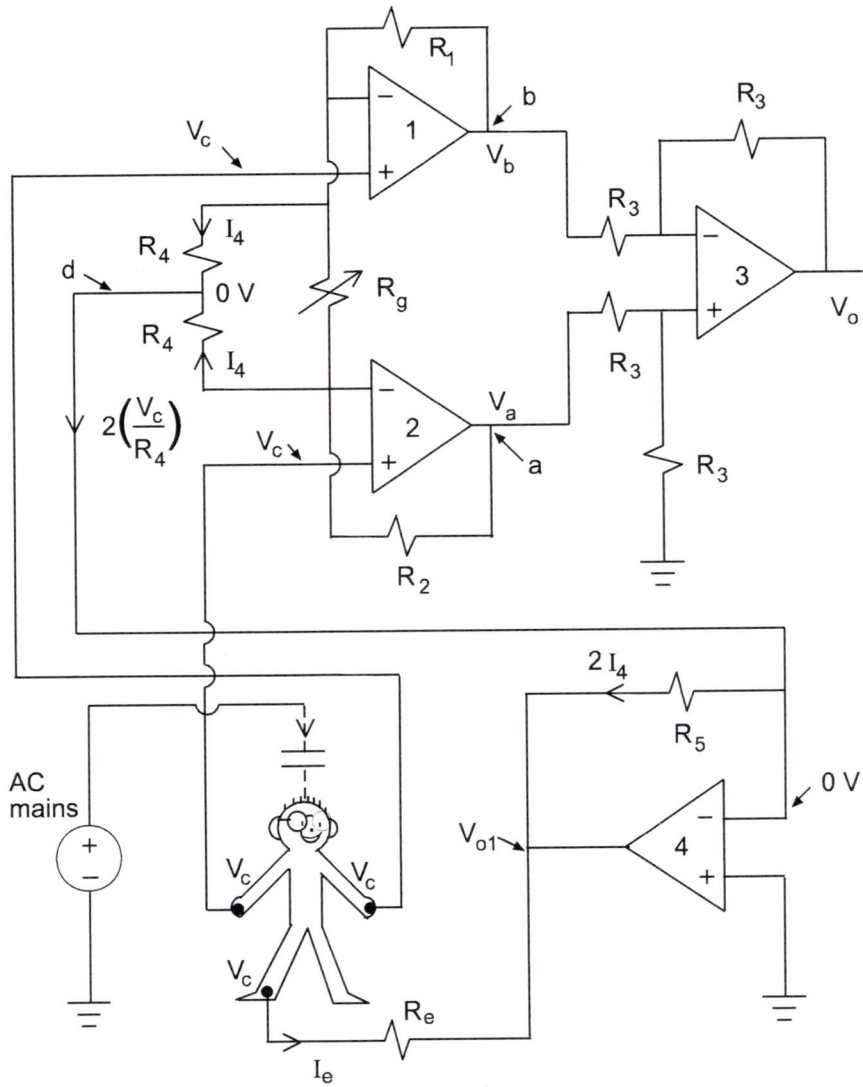

Fig. 13.5 Instrumentation amplifier used with feedback op amp to reduce power line induced common mode body voltage.

left and right arms are connected to the non inverting inputs of op amps 1 and 2, which draw no current from the body, then the inverting inputs of these op amps are also at the common mode

273

body voltage V_c. Then the current through either resistor R_4 is V_c/R_4 and the current that reaches op amp 4 is $2I_4 = 2(V_c/R_4)$. Since this current cannot flow into the op amp it flows through the feedback resistor R_5. Let the voltage at the output of op amp 4 be V_{o1}, and the current flowing from the AC power line through the body and out through the right leg skin electrode (of resistance R_e) I_e. Then Ohm's law gives

$$V_{o1} = -R_5 \left(\frac{2V_c}{R_4} \right)$$ 13.8

$$V_c - V_{o1} = I_e R_e$$ 13.9

Eliminating V_{o1} between equs. 13.8 and 13.9 gives

$$V_c = \frac{I_e R_e}{\left(1 + \dfrac{2R_5}{R_4} \right)}$$ 13.10

Substituting typical values in equ. 13.10 gives

$$V_c = \frac{0.1 \times 10^{-6} \times 50 \times 10^3}{\left(1 + \dfrac{2 \times 1 \times 10^6}{24.9 \times 10^3} \right)}$$ 13.11

$$V_c = 61 \ \mu V$$ 13.12

The driven right leg circuit reduces the common mode body voltage to 61 μV, a reduction of two orders of magnitude, from the 5 mV calculated in the previous section when the instrumentation amplifier was used alone.

13.5 Active bandpass filter

The output V_o of the circuit of Fig. 13.5 is fed into an active filter to increase the gain further by about 200 V/V and to limit the passband from under 10 c/s to below about 3 kc/s. this can be done with the circuit of Fig. 13.6.

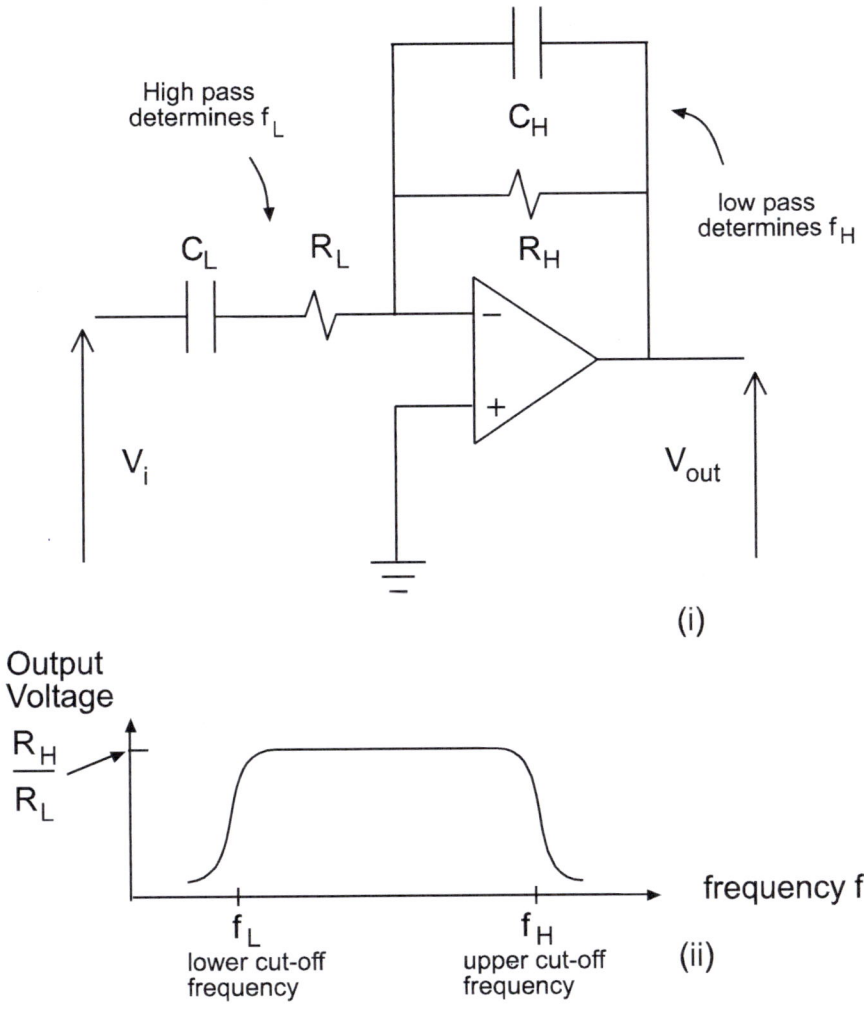

Fig. 13.6 Active filter at final stage of bioamplifier

275

The input stage is the high pass filter with a transfer function of

$$\frac{V_{out}}{V_{in}} = -\frac{R_H}{R_L}\left(\frac{j\omega}{\omega_0 + j\omega}\right)$$

where

$$\omega_0 = \frac{1}{R_L C_L}$$

and the lower cutoff frequency is

$$f_L = \frac{1}{2\pi R_L C_L}$$

The feedback components give the low pass filter with a transfer function of

$$\frac{V_{out}}{V_{in}} = -\frac{R_H}{R_L}\left(\frac{\omega_0}{\omega_0 + j\omega}\right)$$

where

$$\omega_0 = \frac{1}{R_H C_H}$$

and the upper cutoff frequency is

$$f_H = \frac{1}{2\pi R_H C_H}$$

The gain within the passband range is R_H/R_L.

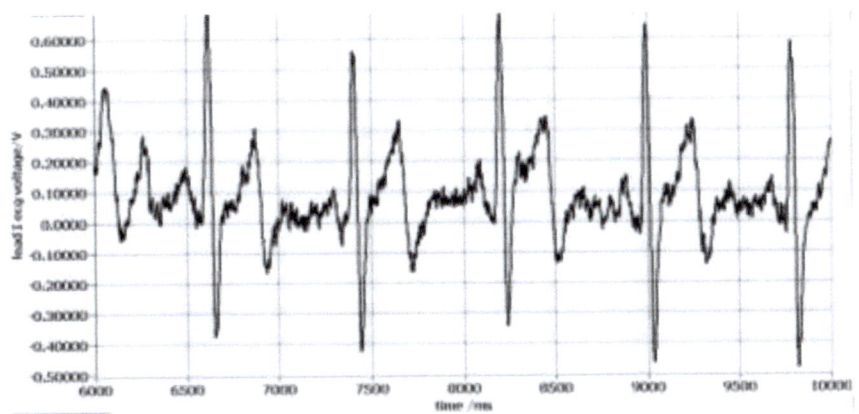

Fig. 13.7 A lead I ECG

Fig. 13.7 shows a lead I electrocardiogram band limited to within 5.8 c/s and 3 kc/s. The gains were 7.7 V/V and 200 V/V for the instrumentation amplifier and the active filter, respectively. The results show that the cardiac signal was in the 65 to 420 µV range.

Appendix

A. Notes on inverting opamp

Figure A1 shows the inverting opamp circuit of Fig. 8.6 wired on a protoboard. The source voltages are not shown connected to the circuit.

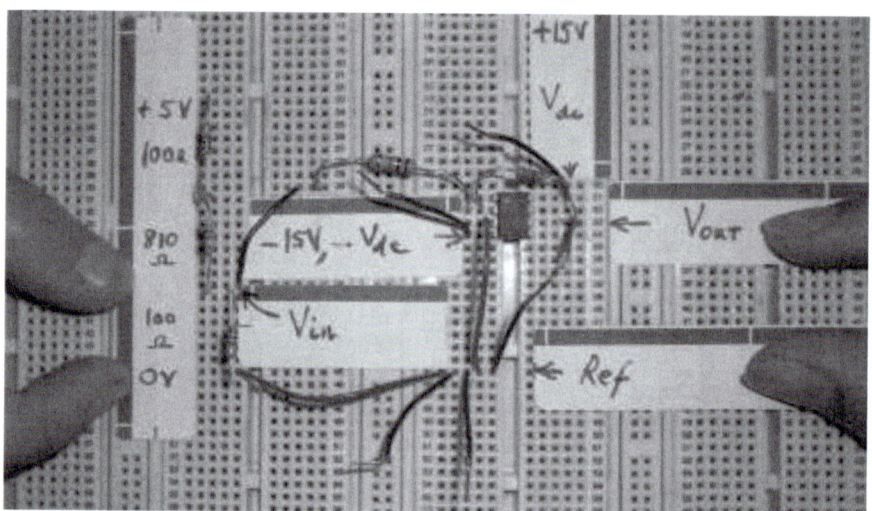

Fig. A1 Inverting opamp circuit wired on a protoboard.

The input voltage would appear between the points indicated on the left. It is approximately 0.5 V that would appear across the 100 ohm resistor, in the three resistor voltage divider circuit, when 5V is applied across the three resistors. The output voltage would then be measured between the two points shown on the right.

Notice that the opamp DIP straddles the trough between two columns of holes for wire contacts. Within a column, holes along the horizontal are electrically connected internally, and can form a single node in a circuit, while there are no electrical connections between holes along vertical lines.

The schematic for the circuit of Fig. A1 appears in Fig. A2.

Fig. A2 Students display the schematic that they used to build the opamp circuit in the Engineering Materials and Electronics course of the Engineering Minor Program at Lehigh University.

Figure A3 shows details of the opamp circuit schematic.

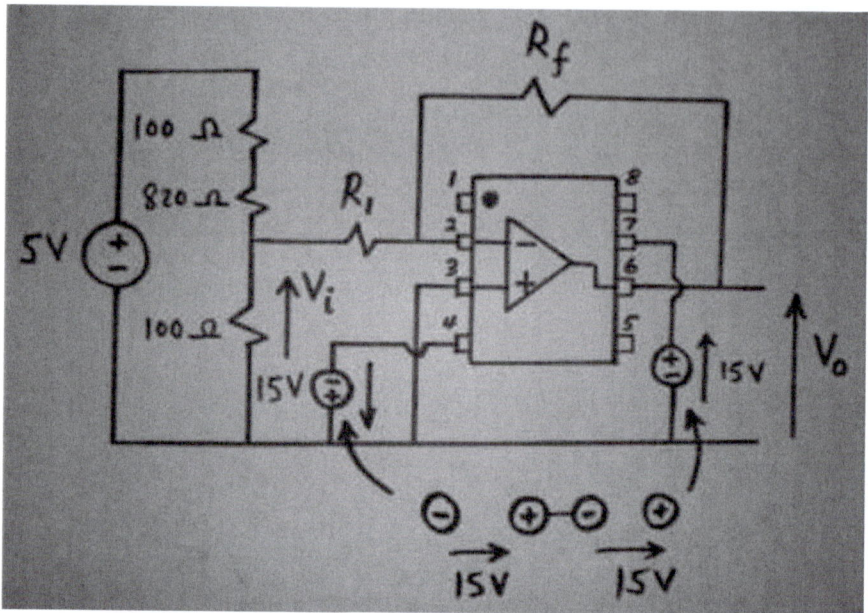

Figure A3 Schematic for opamp experiment. The power supply had two variable sources that could each provide a maximum of 15 volts; these were used for the dc supply voltages. A third source, that is fixed at 5 volt, on the same power supply, was applied across the voltage divider on the left, to provide the 0.5 volts appearing across the 100 ohm resistor, as the input to the opamp. The output voltage could then be varied by choosing different values for R_1 and R_f to decide the gain R_f/R_1.

B. Note on square wave generator

Fig. B1 shows the wiring diagram for the 555 circuit of Fig. 10.6.

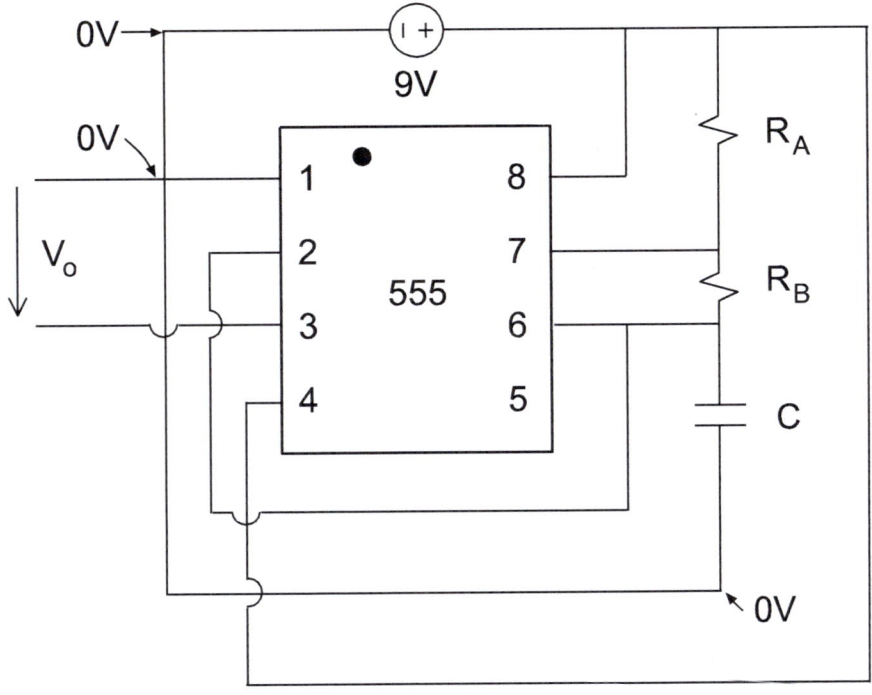

Fig. B1 Wiring diagram for 555 square wave generator.

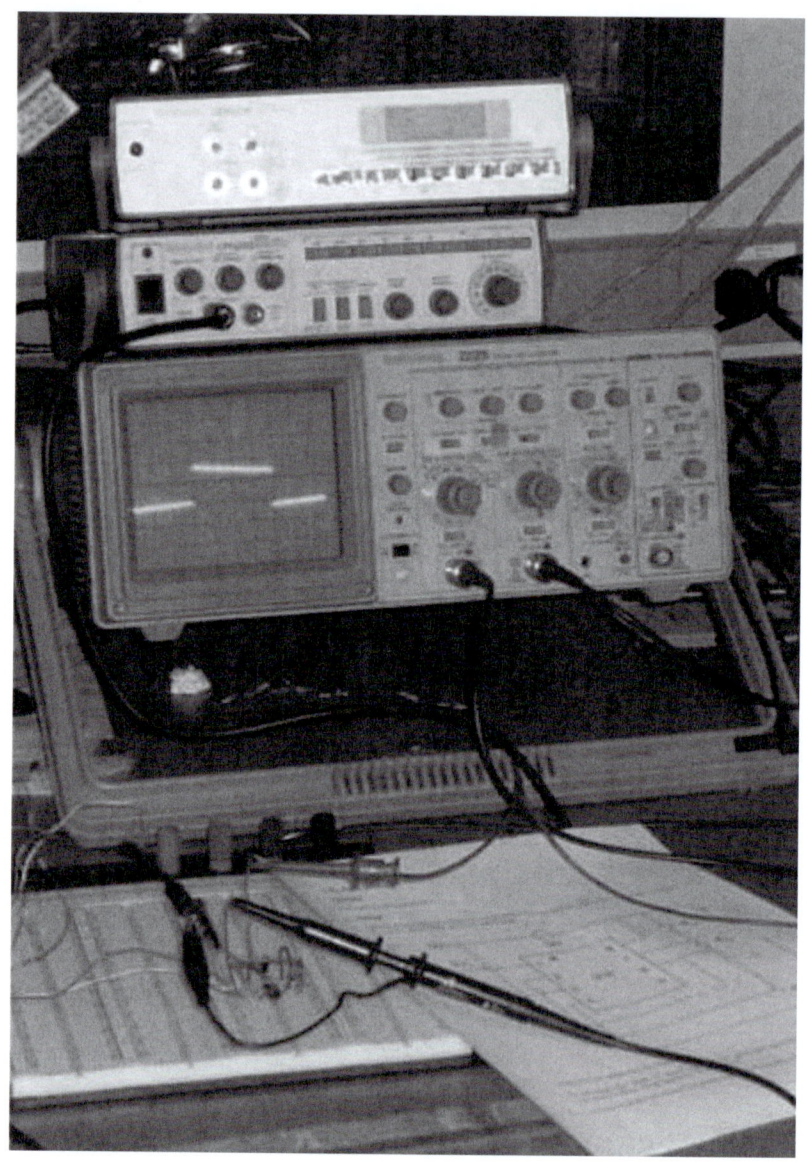

Fig. B2 The square wave generated, by the 555 circuit of Fig. B1, displayed on the oscilloscope screen.

C. Instrumentation amplifier and op amp pinout diagrams

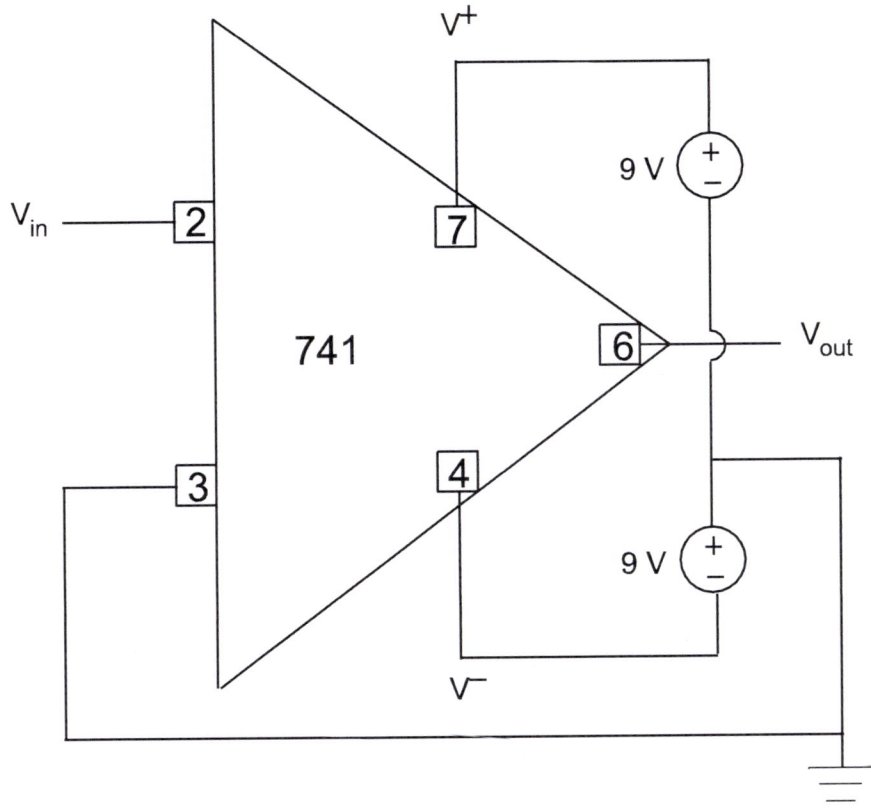

Fig. C1 Pinout diagram for 741

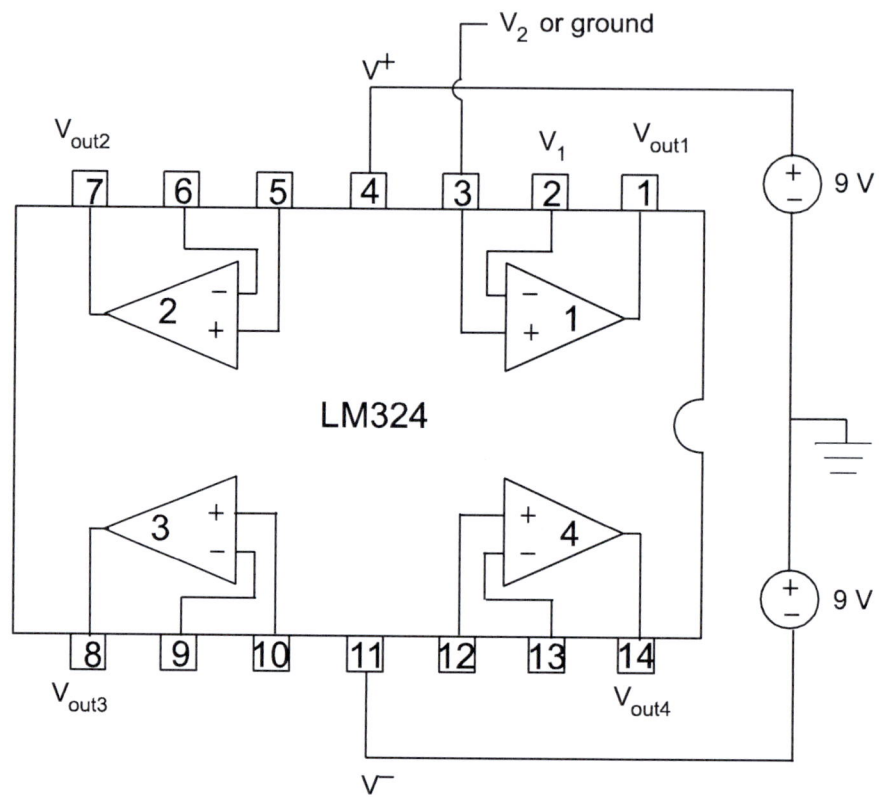

Fig. C2 Pinout diagram for LM324

284

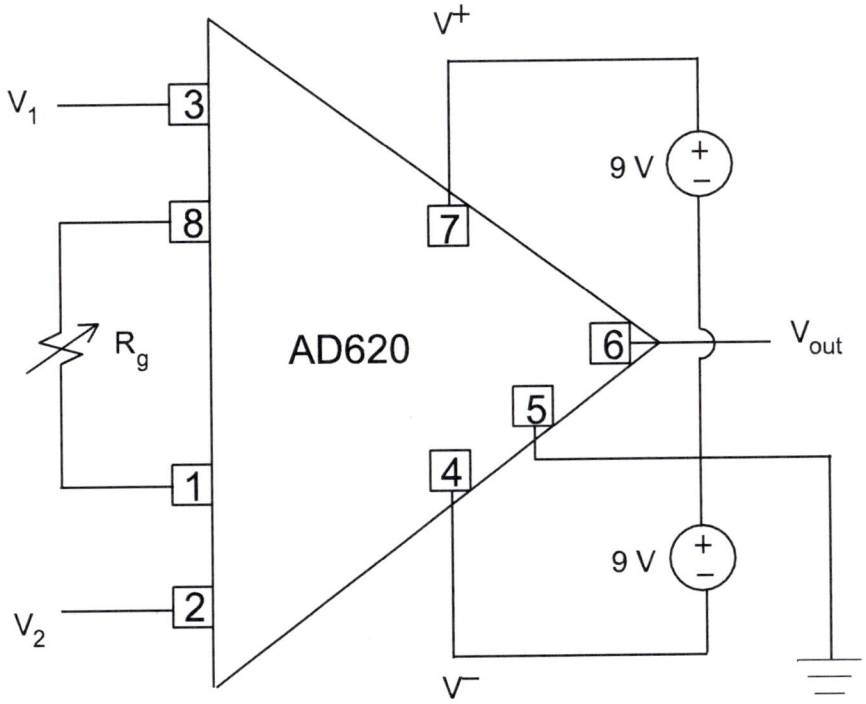

Fig. C3 Pinout diagram for instrumentation amplifier AD620

D. Energy projects photos

D1. Tesla coil experiment

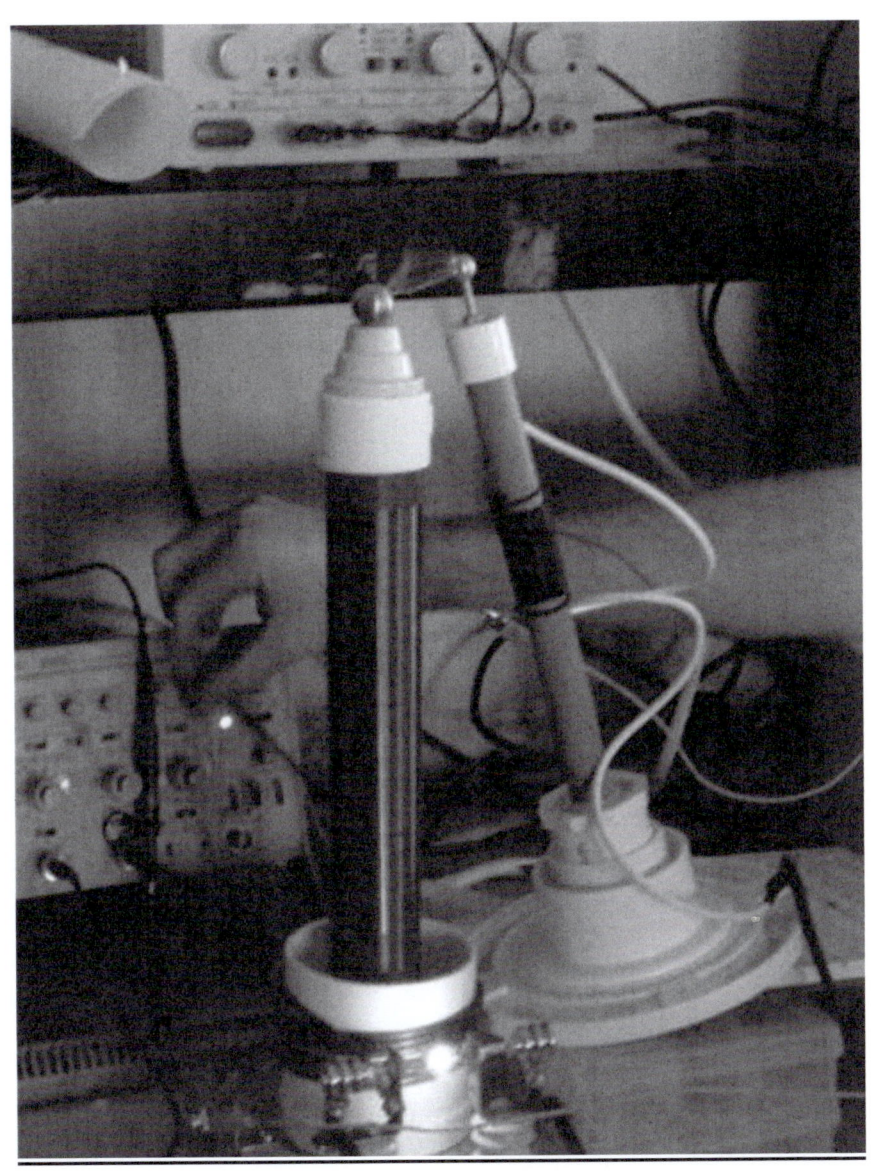

D2. Tesla coil

Index

maximum power transfer · 128
maximum power transfer to load · 57
membrane potential · 265
mesh · 53
mesh current method · 53
minority carriers · 241, 245
mobility · 238
model of actual op amp · 189
monostable multivibrator · 220
motor · 226, 229

N

negative current feedback · 262
negative feedback · 259, 260, 263
negative voltage feedback · 263
nickel cadmium battery · 251
node · 1
node voltage method · 45
noise · 200
non-inverting amplifier · 194
non-inverting opamp · 263
non-inverting terminal of op amp · 187
non-rechargeable batteries · 251
normalized frequency · 205, 207
Norton equivalent · 28
Norton's theorem · 27
n-type semiconductor · 241
Nyquist frequency · 207, 210, 213
Nyquist index · 205, 210, 213

O

odd symmetry · 210
ohm · 4
Ohm's law · 5, 6
ohmic contact · 237
ohm's law · 236
one sided DFT · 205
op amp · 186, 278
opamp 741 pinout diagram · 283
open circuit · 1
open circuit voltage of solar cell · 247
open loop gain · 260
operational amplifier, op amp · 186

overdamped circuit · 170, 174

P

P wave · 267
parallel combination of resistors · 15
passband · 183
passband frequency · 182
period · 90
periodic function · 91
periodic waveform · 90
permeability · 144
permittivity · 61
phase angle · 92
phase plot · 210, 212
phase voltage · 233
phasor power · 123
phasor representation of sinusoidal
    voltage and current sources · 105
phasor representation of sinusoids · 95
photocurrent · 243
photoexcitation · 242
photon · 242
photovoltaics · 236
pn junction · 215, 243
polar representation · 98
polarity · 2, 5
potential difference · 2
power · 5
power consumed by the resistor · 94
power dissipated as heat · 125
power factor · 126, 142
power factor correction · 138
power in three phase circuit · 234, 235
power in transformer · 152
power line induced body voltage · 271
power spectrum · 207
power triangle · 126
primary batteries · 251
primary coil · 142
primary voltage · 148
p-type semiconductor · 241
pulse repetition rate · 222

NOTES